THE HERDSMAN'S BOOK

THE HERDSMAN'S BOOK

by

KENNETH RUSSELL

BSc, NDA (Hons), NDD (Hons)

revision

by

STEPHEN WILLIAMS

CBE, MSc, NDA, FRAgS

FARMING PRESS LTD

WHARFEDALE ROAD, IPSWICH, SUFFOLK

First Edition, 1956

Second Edition, 1962

Third Edition, 1969

Fourth Edition, 1972

Fifth Edition, 1974
2nd Impression
(with amendments) 1976

Sixth (revised) Edition 1978

ISBN 0 85236 089 4

Printed in Great Britain in 11pt on 12pt Times Roman on Longbow Cartridge paper by the Leagrave Press Limited, Luton and London

CONTENTS

LIST OF ILLUSTRATIONS

FOREWORD

by Sir Richard Trehane

THIS revised edition of *The Herdsman's Book* reminds us of the great debt which agriculture owes to a man who devoted his life to serving its best interests.

The enduring value of Kenneth Russell's contribution can be seen on every page. In approaching his subject, an innate sense of stockmanship of the highest order is helped by a keen appreciation of the economics of modern milk and beef production. Russell's reputation was founded on his lucid presentation of scientific knowledge and a practical approach to problems associated with the dairy herd.

This volume, revised by Stephen Williams, is a tribute to Kenneth Russell's great ability to communicate a lifetime's experience in a fashion acceptable and profitable to readers of all ages and experience.

RICHARD TREHANE

Thames Ditton,
Surrey, 1962

PREFACE

(to First Edition)

THIS book has been written in the hope of providing a code of good husbandry for the guidance of those whose job it is to tend our dairy and beef cattle.

Some previous knowledge is assumed on the part of the reader, otherwise to deal with elementary details would make the book unduly long, so I have tried to present existing scientific knowledge in its everyday application.

At the same time I am deeply aware of the personal factor in good stock husbandry; it is a life of dedication with a satisfaction becoming increasingly rare in this world of automation.

My admiration and respect for the "good cow" then is matched only by my regard for the skilled stockman, and the one deserves the other.

KENNETH N. RUSSELL

Askham Bryan,
October, 1956

PREFACE

(*to Sixth Edition*)

HAVING introduced metrication in the 1974 edition, I have now omitted the old Imperial measurements. Metric values have been rounded off. This edition introduces the Metabolisable Energy system of feeding dairy cows.

STEPHEN WILLIAMS

Newtown,
Powys.
April, 1978

CONVERSION FACTORS USED IN TEXT

1 inch	= 2·5400 mm	1 cu inch	= 16·387 cm^3
1 foot	= 0·3048 m	1 cu foot	= 0·0283 m^3
1 yard	= 0·9144 m	1 pint	= 0·5683 litre
1 mile	= 1·6093 km	1 gallon	= 4·5461 litres
1 sq inch	= 6·4516 cm^2	1 ounce	= 28·350 g
1 sq foot	= 0·0929 m^2	1 pound	= 0·4536 kg
1 sq yard	= 0·8361 m^2	1 cwt	= 50·802 kg
1 acre	= 0·404686 hectare	1 ton	= 1·0161 tonne
1 fl oz	= 0·0284 litre		

INTRODUCTION

PROFIT margins in milk production are always changing as a result of continuously rising costs of all the resources that are used in the process.

In response to the economic pressures, producers are always modifying their methods and systems. Most remarkable, the size of herd tends to increase each year as the producer endeavours to maintain his profit margin. At the same time the higher-yielding Friesian replaces cows of other breeds of lesser milk production capabilities.

In response to rising labour costs, labour-economising systems are adopted—milking parlours take the place of shed-milking, cows are loose-yarded rather than tied up in sheds, etc. As herd size increases, often the producer becomes a dairy specialist and ceases to grow cereals, so that he has to buy such bedding straw as he might need and often the cost of doing so persuades him to install cow cubicles.

As land values and rents rise, the producer seeks ways of intensifying grassland use by using the paddock system of grazing and by applying increasing quantities of nitrogenous fertilisers and by irrigation.

Where milk output is elevated by intensification, the economy of using bulk-tank milk storage and collection on the farm soon becomes apparent.

Manure-handling and disposal in these intensive large-scale units still represent a major problem that awaits solution.

There comes a point for decision along the line of intensification when the producer has to weigh carefully whether he should discontinue to breed his own replacements or contract them out to others to raise for him on a financial basis or whether his cattle-breeding efforts reward him more than the increased scale of production would yield to him.

Cessation of keeping young stock on a farm provides considerable scope for increasing cow numbers and, at successive levels of herd size and stocking density, given good management, economies of scale are realised.

Another approach to this aspect of the problem of providing replacements is to manage the cows so that their milking lives are extended and their replacements correspondingly required are the fewer. This means better control of disease in the herd through good milking practice, sound feeding and proper health protection in the form of ideal housing and appropriate vaccinations and therapy.

The cost of concentrates are a major factor in milk production and where suitable acres of grassland are available for intensification it is often profitable to attempt to economise in concentrate usage.

It is important, however, to recognise that to allow yield to suffer is often mistaken economy. Consequently, standards for comparison are meaningless unless the level of milk production per cow and seasonality of production are taken into account.

Records maintained in specialist dairy enterprises are very meaningful and they are essential elements on which to build rational policies and practices. They should cover all aspects of the business—performance, health, labour and land use, feeds and fertilisers used and all accidents and disposals.

By their analysis, the way ahead can be best designed to fit in with new developments that are always being injected into dairy farming.

CHAPTER I

FEEDING THE DAIRY HERD

THE success of the great cattle industry in this country depends on the ability and skill of the men who manage the stock. A good herdsman is essential to the success of any livestock enterprise.

It follows, therefore, that the art and science of cattle husbandry merits careful study by all concerned with producing milk and beef.

The good herdsman, in the first place, needs to appreciate to the full that productivity in his stock depends on proper feeding.

Secondly, he is not expected to be a veterinarian. His job is to prevent disease—and in that aim the proper feeding of stock, as far as it is within his power, is most important.

Nevertheless he must be able to recognise ill-health or unthriftiness at the earliest possible stage and, by skilled attention to hygiene, aim at maintaining a healthy herd.

Thirdly, he needs to understand how to handle animals properly. Sympathetic but firm treatment of stock reaps a handsome dividend.

Such an understanding is born of close observation and interest in one's work. Therein lies the partial truth of the saying "a good herdsman is born not made".

To make a good herdsman better, training is invaluable; and though there is a limit to what can be taught about herdsmanship, it will be my endeavour to provide the "science" in the following pages—beginning with the fundamental job of feeding.

Cattle use food for two main purposes: first to provide the means of maintaining life and normal activity (this is termed the maintenance requirement); secondly to provide for growth or for other forms of production—milk, meat, and the development of the unborn calf.

To be able to express these requirements to a standard pattern, various methods are adopted in different countries. In Britain we commonly use the Metabolisable Energy system which reduces all feedingstuffs to their value expressed in terms of useful energy per kilogram (kg) of dry matter.

This is an excellent means of comparing foods on an energy basis; but, in addition, we need to know the digestible protein content of feedingstuffs and this is given by the term Digestible Crude Protein (DCP) and expressed in terms of grams DCP per kilogram of dry matter, (g.DCP/kg DM).

Feedingstuff Values

At this stage it will be helpful to turn to Appendix 1 on page 163 and to study the table there. It gives the 'useful energy' values of all the common feedingstuffs used in cattle feeding—as Metabolisable Energy (ME) and expressed in terms of megajoules per kilogram of dry matter (MJ/kg/DM). To use this information, of course, it is necessary to know the dry matter content expressed as grams per kilogram (g/kg) and average figures are given in the table. The table also provides the digestable crude protein values for feeds and these are expressed in terms of grams per kilogram of dry matter (g/kg/ DM).

Another point that you will note about the tables is that they classify the feedingstuffs into concentrates, succulents and roughages.

The distinction between them is as follows:

CONCENTRATES are foods of a low moisture content (only 10 per cent to 14 per cent). They are highly digestible, with a low fibre content and a high energy and/or protein value.

Flaked maize and white fishmeal are typical concentrates, as are cakes like linseed and cotton seed.

SUCCULENTS are foods of a high moisture content (over 80 per cent); but the dry matter in them is highly digestible, so that though the starch equivalent per kg of food is low, the energy value in the dry matter is often as high as in concentrated foods.

Root and green crops, grass and silage are typical succulents.

ROUGHAGES are dry foods like concentrates but are much less digestible owing to their higher fibre content. So they are of much lower energy and protein values.

Hay and the cereal straws are examples of roughage foods.

What Cattle Need

The new standards are set out for compiling rations for cattle which, on paper, satisfy the nutritive requirements of cattle. But the stockman must go further than this; he must ascertain certain things by trial. These are:

1. Whether the foods offered are sufficiently palatable to be eaten readily. (Food refusal may not be due to actual lack of appetite.)
2. Whether after each meal each animal's appetite is satisfied. If it is not, then cattle will remain restless and fail to thrive as they should.
3. Whether the "production level" expected is actually being achieved. This involves measurement of production by recording milk yields or by periodical weighing of fattening cattle to check rate of liveweight gain.
4. Whether the ration as a whole is digested without disturbance to the digestive system—as shown, for example, by undue scouring or constipation.

It is most important to realise that feeding standards are a guide to feeding practices only, and the good stockman is prepared to modify his rations to suit varying circumstances. Just a few of the "variables"—which no feeding standard can fully express—are discussed in the following paragraphs.

Appetite is generally reckoned to be 0·02-0·03 kg of dry matter (i.e. food minus water) per kg liveweight per day. But this basis is far from satisfactory in that appetite varies with health, with level of production, with degree of activity, with palatability of foods, as well as with liveweight on which the feeding standards are based.

The conditions under which cattle are kept (whether, for instance, indoors or outwintered) affect the maintenance requirement—which was originally computed for cattle being stall fed.

At pasture the energy expended in movement may add 20 per cent to the maintenance requirement—even more on bare pasture. And in winter, wet cold weather can cause an equally large increase in maintenance of cattle out-of-doors compared with stall-fed cattle.

Furthermore waste in feeding can cause wide differences between foods offered and what is actually eaten.

Variation in the feeding values of foods—particularly home-grown foods like hay and silage (subject to loss of nutrients in the making) can be very wide. The values given in the tables* are for average samples—it is the stockman's business to estimate the feeding value of the foods at his disposal as nearly as he can or to submit samples for analysis and assessment.

With all purchased feedingstuffs a guaranteed analysis has to be given on the invoice under the Fertilisers and Feeding Stuffs Acts. This chemical analysis is usually in the form of percentages of:

1. Protein.
2. Oil—a rich source of energy.
3. Fibre—an indication of the general digestibility of the feedingstuff in that a figure of over 10 per cent would indicate a concentrated food of low digestibility.

In the main, therefore, purchased concentrated foods show much less variation in feeding value than home-grown feedingstuffs.

The Metabolisable Energy System

An Introductory Note—Anticipating the ME System

After a study tour of American farming in 1947, the writer (SW) returned determined to examine in a practical way the principles of Mark Keeney and other dairy farming specialists whose great cattle, high quality roughage, high energy maize silage and liberal feeding methods achieved records of high production and long life.

Our good herd of about 56 Ayrshire cows, in the charge of a wise northcountry herdsman, provided the opportunity. The system followed and developed over many years was carefully recorded and accounted. Liberal feeding of high quality hay,

* See Appendix I, pages 163-165.

with traditional amounts of kale in early winter followed by mangolds, and supplemented by very moderate levels of home-grown cereals and protein feed supplements, were the basis of the feeding system during the winter months.

The amount of concentrated feed was adjusted according to the stage of lactation, the condition (fitness) of the cow, and within the mid-lactation period some regard was paid to yield level.

Monthly weighing of each cow over a two-year period revealed regular cyclical changes in her weight and every member of the herd conformed to the same pattern. Typically, each cow was at a peak weight at calving. She lost 125-150 kg in weight, including the calf, fluids and membranes, in the first 12 weeks of lactation. During the middle and late lactation and in the dry period she gathered condition and weighed 25·4 kg more at the next calving than at the previous one. In the following lactation she dropped to a minimum weight, 12·7 kg higher than her previous minimum. This pattern was generally repeated for the first six lactations gaining an extra 25·4 kg and dropping to a minimum weight, 12·7 kg higher than previously. Each lactation she increased her milking capacity as she increased in weight herself and as she increased her capacity to draw on reserves in early lactation to supplement her food intake. After six lactations, she maintained more or less regular maximum and minimum weights each year. Growth had ceased and her capacity to build reserves had stabilised.

Observations made when this herd was dishorned were significant in relation to milk production. Cows dishorned in the first 12 weeks of lactation suffered a check for two or three days and then fully recovered to previous levels. Those dishorned after the 12th week dropped production and did not recover to their previous levels. It was concluded that the best time to dishorn the herd is very late in lactation or in the dry period.

But the real significance was that the cow makes a valiant and successful effort in that 12-week period to reflect the preparation that went into her feeding regime in the previous six months. Certainly generous feeding is necessary in this period to prevent excessive weight loss, but the idea of feeding to stimulate production seemed spurious. She did not need it nor use it for that purpose; the occasional cow ill-prepared for lactation proved it.

It was clear, too, that the long period of preparation assisted in sustaining yield late in that lactation and was also materially important in the following lactation in achieving and sustaining high production.

It was a very healthy system; cows bred regularly and lived to a great age in that herd and the only heifers introduced were those designed to improve the breeding quality of the herd. In fact, about 80 initial cow families were reduced by selective introduction of home-bred daughters of proven dams to seven cow families in less than 20 years.

Newly calved cows in that herd lost weight in early lactation in every month of the year except the month of May. This seems to suggest that in May the grass is so lactogenic and the cows were in such a state of health that they were able to react to the quality of the grass, and that they had the capacity for intake to ingest enough to meet all their requirements, without drawing on reserves. It is significant in this connection that bulk intake of high quality hay was never restricted during the winter period and thus the rumen capacity was always fully developed.

Though our conclusions were strongly opposed at that time by those who preferred other systems and the theories they believed underlay them, we did not lose faith because what a man does with his own hands he does not doubt. It is a pleasant task, therefore, to introduce here a system which is a scientific refinement of what were clearly fundamental aspects of dairy cow-feeding and management as we then were led to believe by practical experience, careful observation and accurate records.

The Principles of the ME System

The new system, known as the Metabolisable Energy (ME) System, of calculating the rations for dairy cows takes into account the naturalness for cows to make use of reserves in the body in early lactation and also provides for the rebuilding of those reserves. It also provides for the natural growth during the early lactations of the immature cow.

It is now felt that excessive weight loss in early lactation should be avoided, not to exceed 0·5 kg per day. Our experience supports this. In this same herd probably milk fat levels suffered a little and perhaps yields could have been raised over the

950-1000 gallon level which, however, was considered satisfactory at that time from low-cost rations.

The new system recognises that the cow's appetite (intake of feed) increases as she grows heavier and at the times of high production and that it is desirable reasonably to satisfy her demands. Indeed, it is probable that our modern Friesian cows have the capacity to grow into much larger animals (as our Ayrshires did) on such a regime than has hitherto been realised on controlled intake feeding. Of course, this emphasises the need for high quality roughages, especially during very high yield periods. In fact, it is possible that already the liveweight of many cows exceeds current estimates in many herds, and maintenance requirements are under-estimated.

In general, the former attitude of feeding for yield on a day-to-day basis has given way to feeding for a whole current lactation and the next. Indeed, we shall have to bear in mind when considering whole diet feeding the need to take into account the nutritional needs of cows over their whole lifetime because apparently the appetite and utilisation of the diet seem to be stimulated, and maturity is hastened by liberal access to properly designed feeds.

The old standards for meeting the needs of the foetus during pregnancy have been updated and the period of provision has been extended.

Groups

Before attempting to explain in detail ration formulation for a herd it is desirable to give some consideration to the most convenient groupings of the cows, whether two, three or four groups, especially when whole diet feeding is practised. Three groups may be based on calving dates—Early, Middle and Late Lactation groups—and a fourth group may be dry cows and those in late pregnancy, collected together because they are expected to calve next time more or less as a group. First-calf heifers could constitute a satisfactory group.

An Exercise in Anticipation

It is necessary to anticipate the daily production performance of cows in the herd, and necessarily, of each group of cows. Estimates may be based on previous performance, that is to say, in earlier lactations of the individuals or of a comparable group in the previous year or of the same group in the previous period of their current lactation.

Alternatively, peak yield may be estimated by dividing the previous 305-day lactation by a factor, probably 200 or whatever factor is appropriate as experienced in the herd. Another way—peak yield may be calculated from the yield recorded at the end of the second week of lactation by multiplying that by a factor established by reference to previous experience in the herd. The peak may be reached in the sixth week or possibly, when the system has operated fully, the 10th week.

After peak yield has been sustained for some weeks, the yield can be expected to fall by 2½ per cent per week. In this way, it is helpful to create a graph showing the rise and fall of the lactation curve, plotting yield per day against month of lactation. The groups are shown on the time scale and the production target for each group is readily identified.

Formulation of Rations and Monitoring the Performance

The whole diets, whether the ingredients are all mixed together or not, are then formulated for the groups and the calculations are explained in the following pages.

The cows are thus provided with the correct amounts of energy and protein etc. in their anticipated feed intake—all based on their expected yields and their expected appetites. Consequently, there is a constant need to monitor the herd as a whole, and also groups and individual cows, to confirm that anticipated performance in terms of yield and the condition of the cows are being registered. If they are not, then expense is being wasted and the target yield should be lowered, foodstuffs reassessed and the formulation readjusted accordingly.

The Metric System and Dry Matter Basis

It is recognised that our unfamiliarity with metric measurements leads to confusion, and it is regrettable that it is difficult to keep standards in mind expressed in the new metric terms. Adding somewhat to the confusion is the need to express ingredients in feedstuffs, etc. in terms of grams per kilogram (g/kg) instead of using percentages with which everyone is familiar. Further, adding to these complications, in the new

Metabolisable Energy system all measurements of feed ingredients are expressed as grams per kilogram of dry matter (g/kg DM). Consequently, it is important always to think in terms of the dry matter of feedstuffs and to be able to convert the dry matter weights to gross feed weights and vice-versa—and this needs practice.

The ME System

The ME feeding system has been described in scientific detail in Technical Bulletin 33 published in 1975 by the Ministry of Agriculture, and in 1976 ADAS issued an Advisory Paper No. 11, 2nd ed, which extracted and extended important details giving nutrient allowances for dairy cows. Both these information sources are acknowledged, and readers who want to know more about the fundamental principles should refer to them.

The Metabolisable Energy system is the most recent basis of calculating the feed allowances for dairy cows. It is gradually replacing the Starch Equivalent system. Energy requirements are measured in terms of metabolisable energy (ME). The appropriate unit is the megajoule (MJ) or 1 million joules. Metabolisable Energy values are expressed as megajoules per kilogram of dry matter, MJ/kg DM.

As mentioned earlier, tables which give the dry matter content in terms of grams per kilogram of common dairy foods, also the metabolisable energy (ME) in the dry matter MJ/kg DM, and the Digestible Crude Protein, DCP g/kg DM, have been provided for reference in Appendix 1 on page 163.

The 'maintenance only' allowance of ME for dairy cows in the first seven months of lactation increases with increasing liveweight:

Liveweight (kg)	300	350	400	450	500	550	600	650	700	750
Maintenance allowance (MJ/cow/day)	36	40	45	49	54	59	63	67	72	77

The allowance of ME for milk production depends on the fat and SNF content of the milk. Typical allowances are:

Breed	*ME (MJ/kg)*
Friesian	4·9
Ayrshire	5·1
Jersey	6·0

For practical purposes, 1 litre of milk weighs about 1 kg.

Combining the information in these two tables is quite a simple process of addition. For example, a 600 kg Friesian cow giving 10 kg of milk is allowed 63+ (10 x 4·9) = 112 MJ/day.

If, however, the cow is losing weight, 0·5 kg/day, then her ME allowance is reduced by 14 MJ. If she is gaining weight, 0·5 kg/day, her ME allowance is increased by 17 MJ/day and pro rata.

As a guide, the rhythmical pattern through the lactation might be as follows:

Stage of lactation	*Approx. week no.*	*Liveweight change (kg/day)*	*Adjustment of energy allowance*
1. Early	0 — 10	− 0·5	− 14 MJ
2. Mid	10 — 20	0	0
3. Late	20 — 40	+ 0·5	+ 17 MJ
4. 2nd & dry period	40 — 52	+ 0·75	+ 25 MJ

During the sixth to ninth months of pregnancy, of course, cows need extra ME. It is convenient to follow the table below in which daily ME allowances are given for maintenance, pregnancy and growth during these last five months before calving again.

Breed	*Liveweight (kg)*	*Daily ME allowance (MJ/head)*
Friesian	600	71 increasing to 83
Ayrshire	500	66 78
Jersey	350	48 60

To clarify the situation, 'maintenance only' energy allowances in combination with weight-change energy allowances are used for cows in the first seven months of lactation and for barren cows until the end of lactation. During the remaining five months it is convenient to ignore these two scales of allowances and to use the table giving allowances for combined maintenance, pregnancy and growth.

Cows which walk long distances to graze or forage should be provided with an extra allowance of about 3 MJ of ME per mile walked (about 250 g barley). Since standing requires more energy than lying, a theoretical extra 3 MJ of ME would compensate for extra eight hours' standing!

It has been recognised for many years that the provision of necessary daily allowances for high-yielding cows involves careful adjustment of the total feed ingredients to ensure that their requirements are provided within the limits of their appetite. While larger cows have proportionately larger maintenance requirements, their production requirements per litre remain unaltered. Appetite expressed in terms of dry matter is usually about 3 per cent of liveweight, which means that the large cows eat more feed. However, in early lactation and also as calving approaches, appetite tends to decrease. Other factors operate depending on palatability, acidity (of silage), high water content (of roots and green crops) and texture (of the feed ingredients).

So it is important to examine the probability that the daily feed prescribed can be happily consumed by the cow and if necessary to make adjustments to the prescription. Fortunately, cows' appetites increase when lactation peaks and when feed quality and digestibility are improved, so these are helpful features. Merely as a guide, appetite figures are available:

Breed	*Liveweight (kg)*	*Appetite (kg DM/day)*
Friesian	600	18
Ayrshire	500	15
Jersey	350	11

Protein Allowances for Dairy Cows

The energy allowances for dairy cows having been considered, it is necessary also to ensure that the day's diet provides adequate protein for their maintenance and production.

Digestible crude protein (DCP g) allowances for dairy cows are given (1) for 0-7 months post calving—for maintenance only and (2) from 8 months calved to next calving.

Breed	*Jersey*	*Ayrshire*	*Friesian*	
Liveweight (kg)	350	500	600	700
0 – 7 months (M only) DCP g/day	225	300	345	385
8 months – next calving (M+preg) DCP g/day	340	430	495	565

DCP (g) allowances for milk production per kg are given below:

Breed	*Fat g/kg*	*SNF g/kg*	*DCP (g) allowance per kg (typical milk)*
Friesian	36	86	48
Ayrshire	38	87	53
Jersey	49	91	63

Supplementary Mineral Feeding

Elements most likely to be deficient in dairy cows are phosphorus (particularly in high-yielding cows) and magnesium, especially in spring and autumn. Manganese and cobalt deficiencies appear on certain soil types and iodine and copper are lacking in some situations. Salt feeding is advisable where stock are kept intensively and where potash fertiliser usage is heavy. Minerals are best incorporated in the ration, though the provision of salt licks meet special cases.

Feeding Rate according to Yield

In the strictest sense, this is not really necessary and, in fact, contradicts some principles involved in the ME system. However, for convenience, a dairy compound or mix of food which have an ME of 11 MJ/kg *as fed* can be fed *over the whole lactation* as follows:

for Friesian milk @ 0·44 kg/kg milk
for Ayrshire milk @ 0·46 kg/kg milk
or Jersey milk @ 0·54 kg/kg milk.

The table below, however, reveals a more accurate approach to the complete feeding of a typical Friesian cow and it should be noted that the weight of compound fed per kg milk changes with advancing lactation.

Stage of lactation	*Week No.*	*Milk yield (kg)*	*Compound food as fed daily (kg)*	*Compound food as fed per kg milk (kg)*
1	0–10	15–25	12·2	0·49
2	10–20	25	9·1	0·36
3	20–30	20	7·7	0·39
4	30–40	15	5·7	0·38
5	40–50	10	4·1	0·41

It is important to differentiate between referring to a *food as fed* and a food DM. The former refers to the weight and MJ value of the actual foodstuff which might contain 100 g/kg water and have an MJ value of 11 MJ per kg, whereas the same food referred to as DM contains no water and has an MJ value of 12·5 MJ per kg. In the table, the compound food is assumed to have an energy value as fed 11 MJ/kg and its DM has a value of 12·5 MJ/kg.

Complete Feeding of Dairy Cattle

Reminiscent of the old tub-feeding system of feeding dairy cows during last century in our city dairies, a new development of forage-wagon feeding of the dairy herd is of considerable interest. The principal features of the system are as follows:

1. The complete diet is fed *ad lib* to groups of cows at the same stage of lactation.
2. The highly palatable mixture comprises bulk feed, cereals, protein feed including vitamins and minerals, such feeds as sugar-beet pulp, and perhaps molasses.
3. The ingredients are specially prepared before admixture. Barleys straw and hay are 'tub-ground'. Vitamins and minerals are incorporated in the protein feed before bulk-mixing in the whole diet. Cereals are rolled or ground.
4. The whole diets, each suitable for early- mid- and late-lactation cows and for young stock, are carefully formulated to provide appropriate energy concentration,

daily dry matter intake and daily energy intake and protein content.

5. Diets can be provided in the same way during the summer when the herd is either zero-grazed or at grass.

6. Mechanisation is comprehensive and expensive involving tub-grinder where essential, mixer-feeding trailers incorporating facilities for weighing in ingredients, special hydraulic loading equipment and augers.

7. Building work is necessary to cover the long feed-line and to provide suitably designed and large feed troughs. The feeding area and driveway are concreted.

8. Work is undoubtedly reduced but special skills are required. The performance of the herd must be related to the theoretical potential of the diet. In the interest of economy the diet needs constant monitoring by the adjustment of ingredients to achieve best performance at least diet cost.

The system is capital intensive, labour-efficient and nutritionally sound. For 150-cow herd, capital purchases of buildings and equipment could cost as much as £43,584 or £290 per cow (1977). The annual charge in this instance would amount to £58 per cow per year. However, where some facilities already exist and where 300 cows are involved, the annual charge could be much less, even only £8 per cow.

Practical results have reached good physical standards of dairy farming. Management decisions are simplified. Cow groups are managed as feeding units. Very large herds are most likely to justify the heavy capitalisation involved.

Keeping an eye on the Cows

The dry period. After drying off in fairly good condition, there should be a continuous improvement in the quality of the ration to achieve 0·7 kg gain per day and to bring the cow into a strong condition at calving and with appropriate udder development. It should be realised that appetite for bulk fodders tends to decrease as calving approaches because increasing demand for space is made by the developing foetus. Good condition at calving prepares the cow for a high peak yield and a well sustained performance because she can draw on reserves in those crucial weeks when her yield is building up and still be in good enough condition to maintain the high level of production.

The first 10 weeks. This is the period in which to feed the *best* forage with gradual, careful increase in concentrate supply watching the cow's behaviour—breathing, perspiration, nervousness—her eyes and the manure which should be of average consistency. It should be realised that any increase in concentrate supply is likely to reduce forage intake, and so the balance between concentrate and forage in the rumen should be changed gradually. Some high quality hay is a safeguard against many nutritional accidents at this time and this is probably the critical period when performance throughout the whole lactation is designed. If forage available is not of high quality then, and only then, should it be restricted to about 6 kg DM daily to allow for uptake of sufficient concentrates to meet her requirements. High energy concentrates are useful in this period and an adequate protein content is justified to counter the tendency of starchy rations to partition in favour of liveweight gain rather than milk production.

Mid-lactation. After the peak of lactation is passed yields usually decrease about $2\frac{1}{2}$ per cent per week, and feeding levels should be designed to keep to this line of decline. Appetite is now at is maximum and the cow will eat more forage as concentrates supply declines. The cow's liveweight loss should have been halted and held from the 10th to the 20th week, and thereafter feeding should provide for a gain of 0·5 kg per day.

The drying-off period. Steady weight gain, 0·5 kg/day, can be provided for from an entirely forage diet *ad lib* economising thereby in ration cost and helping lactation to come to an easy close. The quality of the forage will determine the amount eaten and the production level. It may be necessary to use less than best quality to assist drying off for a short period.

Quality of Bulk Feeds

The herdsman's next job is to estimate as far as possible the quality of his bulk feeds available. High quality hay and silage will enable some saving in supplementary feeding of concentrates to be achieved, and maintain milk yields at a higher level throughout the winter.

Indications of high quality in bulk feeds are:

HAY:	Absence of mould Green or yellowish-green colour	Well made
	Soft texture—little loss of leaf No seed formed	Early cut
	Analysis—fibre content not more than 28 per cent and protein content 8-12 per cent in drymatter	
SILAGE:	Greenish-yellow colour Sharp acidic smell	Well made
	Few flowering heads	Early cut
	Analysis—drymatter not less than 22 per cent and protein 13-15 per cent in drymatter	

Chemical analysis (available through the ADAS) is a useful guide in the preliminary assessment of hay and silage quality.

General Feeding Routines

Cowshed Housing

With cows individually tied in cowsheds it is possible to vary feeding according to stage of lactation, and this is facilitated by tying up the cows in groups as they calve during the autumn.

The freshly-calved cows and cows in milk up to the fourth month should have the best quality bulk feeds available, whereas the stale milkers can be fed poorer quality bulk feeds (if any) and in greater quantity.

Such a feeding system makes the best use of bulk feeds with minimum usage of concentrates.

A further modification in feeding programmes is necessary where cows are being fed for very high yields, in association with three times milking (as in milking trials). The concentration of the diet is carried a stage further by reducing bulk feeds to the level required for satisfactory rumination and feeding a wholly concentrate diet thereafter.

Dried sugarbeet pulp, fed after soaking to a crumbly consistency, or wet brewers' grains are commonly used as the succulent feeds. Hay should not be reduced below the level of 9 g hay per kg liveweight of the cow to maintain a satisfactory physical texture to the contents of the rumen or first stomach.

The daily feeding programme would be:

6.0 a.m.	concs hay	1st milking
10.0 a.m.	beet pulp or wet grains	
1.0 p.m.	concs hay	2nd milking

4.0 p.m.	beet pulp (fed soaked) or wet grains	
9.0 p.m.	concs hay	3rd milking

With increasing costs of labour this system of feeding and milking is not of *general* application in commercial milk production today.

Yarding and Self-feeding

In successful self-feeding and loose-housing (yarding) systems, the quality of the silage available is of the greatest importance, as the cow herself governs her feed intake.

Self-feeding of silage is, in contrast to controlled hand feeding in a cowshed, an inflexible system offering much less scope to the herdsman to modify the cow's diet by changing the proportion of feeds available to the cow. Beyond moving the electric fence, electrified pipe or feeding barriers, and cleaning up 'slurry' at the silage face, the herdsman has no routine feeding programme to follow.

He should, however, try to calculate the general level of silage consumption per cow in order to know what level of supplementary feeding is required.

On unrestricted 24-hour access to a silage face allowing 150-230 mm of feeding space per cow (according to breed), a cow will eat approximately 4·5 kg of good quality lacerated silage per 50 kg of her liveweight in a total feeding time of about 4 hours. Thus a 559 kg cow would eat around 51 kg of silage per day, or 0·76 m^3 of silage in nine days.

With restricted access, allowing 0·76 m of feeding space per cow so that all the cows can feed at once, silage consumption can be reckoned to be at the rate of 9-13 kg *per hour*, again depending on size of cow.

With poor quality or badly-fermented silage consumption falls much below these levels and dry matter intake, particularly with wet silage (below 20 per cent dry matter) is correspondingly reduced to a level probably meeting no more than maintenance requirements. Some knowledge of the quality of the silage and the amount being eaten is obviously of vital importance to the herdsman.

The winter months of 1973/74 presented a dilemma to intensive high level milk producers who were confronted with high cost feedstuffs without adequate milk prices necessary to achieve overall profitability. In the circumstances science can suggest no magic formula to resolve the economic problem. Feeding for lower yields produces less milk; cutting down feed drastically produces less milk and then barren cows and cows unable to respond to grazing in the spring.

Management Points

No matter what the system of feeding and housing employed, a herdsman should pay regard to the general health and well-being of his cows. The following management points deserve special emphasis:

1. Observe regular feeding times and allow adequate periods of rest between meals—regular rumination is most important. Give the cows all possible comfort.

2. Avoid over-feeding in any one meal—a cow should leave her feeding trough clean. Concentrates should be fed first, followed by hay and then succulent foods, in the course of any one 'meal'. Litter down after feeding.

3. Watch the condition of the cows' dung—a mildly laxative condition is desirable. Succulent foods tend to promote laxativeness, roughage foods are more costive; oil cakes tend to laxativeness, as does molasses or bran fed as a mash. Beef cattle should be firm in the dung.

4. Tenderness in the feet is often associated with excess protein consumption. High protein foods in the concentrate ration should be balanced with cereal foods (see pages 76-77) in relation to the quality of the bulk feeds available.

5. Loss of appetite may be due to—overfeeding; unpalatable foods (e.g. dirty roots, mouldy hay, or 'rotten' silage); a derangement of the digestive processes, as with acetonaemia; ill-health, e.g. with pneumonia or inflammation of the genital organs. Try to ascertain the cause before offering further food, and if in doubt call in veterinary advice (see chapter VIII).

6. Feeding influences the compositional quality of milk within the limits set by the breeding (hereditary make-up) of the cows. Butterfat production is low on diets deficient

in long fibre—in winter this could be due to too little roughage being fed, or overfeeding with concentrates. Underfeeding of concentrates where the diet is generally deficient in energy, as with self-fed silage, would result in low snf, unless the silage is suitably supplemented with cereals (i.e. energy providing foods).

With beef cattle the same general principles apply. To fatten quickly with high conversion rates, beef cattle need to be well fed (though not overfed), comfortably housed, and disturbed as little as possible during the day.

With all yarded cattle keep a watch for the persistent bullies which can so upset any feeding regime. It may well be worth while to yard such cattle separately.

The SE/PE System of Feeding Dairy Cows

While it is important to give the essential details of the new ME system, there will be some who are still interested to understand the operation of the old Starch Equivalent system which served successfully for 50 years. This system can be appreciated by referring to table given below, (SE. Starch Equivalent; PE. Protein Equivalent):

Breed	*Average liveweight (kg)*	*Maintenance req't*		%	*Production req't for 1kg milk*	
		as SE (kg)	*as PE (kg)*	butterfat	*as SE (kg)*	*as PE (kg)*
Friesian	560	3·2	0·34	3·5-4·0	0·25	0·05
Ayrshire	460	2·7	0·29	3·5-4·0	0·25	0·05
Jersey	380	2·5	0·23	5·0-5·5	0·32	0·08

For the benefit of simplification, Kenneth Russell converted Starch Equivalent values into terms of Hay Equivalents, whereby:

1 kg average hay = 3 kg silage
= 4 kg kale, beet tops or swedes
= 5 kg mangolds
= 2 kg barley or oatstraw
= 0·5 kg dairy cake

The cow's requirement was also expressed in the same terms:

Breed	*Liveweight (kg)*	*Requirement*	
		Daily Maintenance	*Production of 1 kg milk*
Friesian	560	9·0 kg hay equiv	0·35 kg dairy cake
Ayrshire	460	7·0 kg hay equiv	
Jersey	380	5·4 kg hay equiv	0·50 kg dairy cake

CHAPTER II

PRINCIPLES OF GRAZING GRASS AND KALE

IN the previous chapter I dealt principally with the application of feeding standards to rations which were, in the main, hand-fed.

Now comes the problem of applying them—as far as is possible—to grazing of grass in summer and to green crops such as kale in winter.

It is true that when cattle are free to graze their feeding is much less under the herdsman's control than when they are stall or yard-fed in winter. Nevertheless the basic principles of providing a ration that is adequate in all respects must be applied so far as the herdsman is able to do so.

The problems—and their solutions—will be better understood if we first study some of the facts we know about grass as a food:

1. Grass is most productive when grazed before any flowering heads appear. Once grasses run to ear feeding value falls rapidly—with decline in digestibility.

2. Grasses and clovers vary in their productive ability, and in their seasonality of growth. Modern leys are sown with highly productive strains, and our best permanent pastures contain a high proportion of productive grasses such as the ryegrasses, timothy, cocksfoot and so on. These grasses have a high tillering capacity and, with clovers, are able to establish a dense sward.

3. On dense swards cattle are able to consume a full diet of herbage with the minimum of movement and exertion.

Sparse or low-density swards may fail to satisfy appetite, and grazing them may entail considerable expenditure of energy by the cows. This will mean an increased maintenance requirement and lowered production.

4. Palatability of the herbage is also important.

Cattle will not graze near their own droppings, so fresh, clean pastures are essential for high production of meat or milk. A sward that fails to give contented grazing stock lacks quality.

Making Best Use of Grass

Higher production of grass by better selection of seeds, better ley establishment and better manuring, has made great strides in recent years. Now the twin problem of better utilisation is receiving much attention from research workers.

Pasture shows a great seasonal variation in yield as well as in feeding value. In the growth flush of early summer, production and feeding values reach their peak. Then comes the "midsummer drop" as growth (due, partly, to drought) falls off, and the herbage becomes more stemmy. Then follows a less marked autumn flush of grass, soon to be cut short by the onset of winter.

Five Problems to Face

The problems of grass utilisation thus involve:

1. A stocking rate on any given area high enough to ensure that the grass is eaten off quickly with minimum wastage of herbage, particularly during the 'flush' season of growth.

Control of the area being grazed is achieved by the use of 'strip' grazing with an electric fence or 'paddock' grazing with more permanent fences. Both systems of grazing control depend on a sufficient 'density' of stocking to achieve full utilisation of available herbage—even so, utilisation rarely exceeds 60 per cent of the actual grass available owing to soiling with dung and urine and treading.

2. To ensure that the most valuable grass is eaten by the most productive livestock.

Dairy cows in full milk or fattening cattle should receive

priority, as should young calves which can be run "forward" of the cows—a practice well established in New Zealand. Store stock, such as bulling heifers and cows in late lactation, can act as followers.

3. To know when and how to supplement grazing to maintain growth rate or milk yields at the various periods during the grazing season, according to the feeding value (growth stage) of the grass and its availability as affected by seasonal factors such as temperature and rainfall.

4. To safeguard the health of cattle on highly productive swards by reducing the risk of bloat (see p. 123), lactation tetany (see p. 130) or parasitic infection (see p. 125), through supplementary feeding or control of grazing.

5. To alternate grazing with cutting where possible, so that any surplus grass is removed to preserve the clover/grass ratio in a sward.

When grasses are allowed to run to ear clover tends to be depressed—this also happens when nitrogenous fertilisers are used without rapid and close grazing or cutting. As cattle refuse to graze near their own droppings (as faeces or urine), cutting also freshens up a pasture and subsequent grazing is more uniform.

Management, therefore, can do much to maintain swards at a high level of palatability.

Ways to Graze

There are four methods of controlling grazing in common use:

"*Rotational*" *grazing*. This involves the grazing in sequence of a series of watered paddocks at a stocking rate of about 25 cows/hectare over a grazing period of 4-5 days. Some 5-6 paddocks are usually required in early spring and summer, with two additional paddocks in reserve to meet the 'summer' gap in grass growth.

Intensive paddock rotational grazing. From 10 to 21 small paddocks, 0·5 to 1·5 ha are grazed in rotation at a stocking rate of 15-25 cows per hectare over a grazing period of one to two days. Heavy dressings of nitrogenous fertiliser are commonly applied throughout the grazing season and surplus grass is conserved as hay or silage in appropriate paddocks in early summer. This system demands considerable skill in grass management and cowmanship, and care should taken by appropriate supplementary feeding of the herd to ensure that the progress around the paddocks does not accelerate as season advances. Too frequent defoliation of the herbage—overgrazing—reduces the yield of grass products and wears out the sward. The damage appears in the form of its invasion by poor herbage species and poorer response to fertiliser use in the future. Soil texture is damaged eventually if overgrazing is persisted in. The sward becomes so thin and the roots system so eroded that in wet weather a 'toe-pan' of puddled clay is formed which impedes surface water penetration, soil air exchange and root penetration.

Strip grazing. An electric fence is used to give a stocking rate per 'strip' of about 200 cows/hectare. In large fields a back fence is used to prevent grazing of the grass as it begins to grow again.

At the same 'stocking rate' there is little difference between field rotation and strip grazing systems—choice is largely a matter of field layout and convenience of access to water. With irregular-shaped fields the rotational system is easier to manage —fences can be erected in the winter and so save the task of moving the fence daily in spring and summer when field work may be pressing.

Set Stocking. This is the oldest system and is usually practised with mixed grazing stock, e.g. beef cattle with sheep.

The system works well if the grazing capacity of the field is known beforehand. With dairy cows, however, it is the least satisfactory method—adjustment of numbers to the grazing capacity of a field is virtually impossible with a 'herd' of cows, the herd being an inflexible grazing unit.

SUPPLEMENTARY FEEDING ON GRASS

Where supplementary feeding is considered necessary or when grass consumption is better limited to below appetite (as with early spring grazing), the electric fence enables this to be done, whereas on rotational grazing a time limit for grazing is imposed.

Supplementary feeding at pasture—except in periods of drought—is designed to 'balance' the high protein content of grass with starchy foods such as crushed oats or barley or dried beet pulp. It will also to some extent alleviate the risk of metabolic disorders such as bloat or lactation tetany.

Furthermore, in the early stages of grazing heavily-fertilised leys some dry roughage foods such as hay or oat straw should be fed with limited grazing. This should be made a regular practice with dairy cows on first turning-out in the spring.

Beef cattle do not need much in the way of supplementary feeding. A little roughage when first turned out (say 2-3 kg hay daily for a week or two) should be adequate.

No further supplementation should be needed until pastures fall off in feeding value after the autumn (September) flush, then cereals or dried beet pulp may be necessary; alternatively, beet tops, apple pomace or turnips can be fed at pasture to finish the cattle without yarding.

Common levels of feeding to assist in finishing beef cattle would be up to 3 kg a head daily of oats, barley or beet pulp, up to 18 kg of beet tops or turnips and from 12 kg of apple pomace or wet beet pulp.

Supplementary feeding of dairy cows when at grass is outlined in the table on the next page. The feeding of concentrates at levels of milk yield below those shown is unlikely to be economic at present milk prices and cost of concentrated food.

Three Grazing Hints

There are three special points to note in connection with the suggested scheme.

First, bear in mind the need for adjusting the grass consumption per cow by increasing the area grazed or the time allowed for grazing, according to the rate of decline recorded in milk yields.

Next, feeding some roughage when cows are on young, highly-nutritive pasture is also commendable for its effect on maintaining the butterfat percentage in milk. This often shows a marked decline in the early part of the grazing season.

Thirdly, when pastures fail during drought conditions, the supplementary feeding of the cows will have to be on a much more liberal scale if milk yields are to be maintained.

Silage offers an excellent standby in this respect; so does lucerne, which can be grazed quite well behind an electric fence.

If blowing is feared, the crop should be mown and wilted for 24 hours before grazing.

Value of Autumn Grass

The milk-producing value of autumn grass is, except in the case of maiden seeds sown the same year, below that of similar grass in the spring flush.

Feeding Cows at Grass

Time of year	*Growth stage of grass*	*Roughage, etc.*	*Concentrates for production*
Early spring (April)	10-15 cm high Young and leafy (Grazing controlled 4 hours daily—half appetite)	Hay 3 kg	*Cereals at 0·4 kg per litre over 13 litres
Early summer (May)	20-25 cm high Pre-flowering stage (Grazing to appetite)	None	Cereals at 0·4 kg per litre over 18 litres
Summer (June)	Flowering stage (Grazing to appetite)	None	Balanced concentrates for yields over 11 litres
Mid-summer (July)	Flowering stage (Grazing to appetite)	None	Balanced concentrates for yields of over 9 litres
Late summer (Aug)	Aftermath or green fodder in times of drought (Grazing to appetite)	None	Balanced concentrates for yields over 4·5 litres
Early autumn (Sept)	Young aftermath or maiden seeds (Grazing to appetite)	3 kg hay or 13kg kale or green catch crops	Balanced concentrates for yields over 9 litres

*Mineral mixture added, rich in magnesium.

Notes:

1. On early spring grazing provide a mineral mixture rich in magnesium in the cereal supplement, as a precaution against lactation tetany or hypomagnesaemia.
2. Appetite level on grazing is usually achieved by allowing:
   With strip grazing—50 m^2 per cow per day
   With paddock grazing—400 m^2 per cow per week
3. Productivity of grazing:
   1st class grazing 100 cow days/hectare
   2nd class grazing 75 cow days/hectare
   3rd class grazing 50 cow days/hectare

This lower feeding value is generally appreciated by herdsmen, but the scientific explanation is less clear. I believe that the higher fibre content of autumn grass and lower oestrogen content are possible reasons.

In practice, the production value of autumn grass is at least 5 litres per cow less per day than similar grazing in spring. Failure to appreciate this fact, and thus be too late in providing extra supplementary food for the cows, is often the explanation of a catastrophic fall in the milk yield of cows in early winter, particularly after a flush of autumn grass.

It is very easy to over-estimate the value of autumn grass to heavy-milking cows and so to cause a drop in the production which cannot be regained. So start winter feeding in good time.

Advantage of Maiden Seeds

There seems to be a definite advantage in maiden seeds for autumn grazing. They appear to have a higher milk-producing value than older leys or permanent pasture; so where it is possible to have a small acreage available each year in the course of the rotation, it will prove particularly valuable for this purpose.

Other supplementary grazing crops will also be useful at this time of year.

Those usually grown are rape, kale, and ryegrass and trefoil. Assuming that the pasture being grazed provides for maintenance purposes (which is probably true in most parts of the country until late September) about 3 kg of one of these green crops will provide for 4·5 litres of milk.

Electric Fencing Guide

At this stage it may be helpful to discuss some practical points in the use of electric fences, not only in relation to grazing grass but for rationing green-crop grazing.

Follow the maker's instructions carefully in erection. Ensure a tight wire, effectively insulated and free of any possible earthing by contact with the crop being folded off.

Run the fence so as to include a water trough, or to give access to the trough from the area being grazed. Particularly when on rape or kale, cattle show a marked desire to drink before the appetite level of consumption is reached.

The following table indicates the acreage requirement at two levels of feeding in respect of the crops shown:

	Hectares per 20 cows per day at	
Yield per hectare tonnes	*Half appetite level* (1 feeding period)	*Full appetite level* (2 feeding periods)
9.9 tonnes Grass	0·05	0·10
19.8 tonnes Rape	0·025	0·05
39.5 tonnes Kale	0·012	0·025

In grazing large fields of grass, where the time required to graze the whole area will exceed a week, it is advisable to use a back fence. This prevents defoliating the sward as it recovers from the previous grazing.

All-Year-Round Grazing

In this chapter, the importance of proper management in the utilisation of increased productivity obtainable from new leys has been stressed.

Grass is the cheapest source of food for cattle and today every effort should be made to provide for as long a grazing season as possible by the use of fertilisers on special leys and catch crops to provide grazing, as far as soil conditions permit, over all periods of the year.

Here is the kind of programme that should be considered:

Early spring grazing (*early bite*). From autumn-sown cereals and Italian ryegrass on sheltered, well-drained fields, top-dressed in February with nitrogen fertiliser.

Early summer grazing. From perennial ryegrass-dominant leys.

Main summer grazing. From meadow fescue/timothy and cocksfoot-dominant leys, and from lucerne.

Autumn grazing. From ryegrass-dominant leys after a mid-summer rest (following nitrogenous fertiliser application).

Winter grazing (*foggage*). From cocksfoot-dominant leys given an autumn rest following nitrogenous fertiliser application in early September.

Forage crops like kale and rape, sown in midsummer or early autumn, are most valuable winter grazing crops where conditions allow.

To the herdsman I would offer this advice: watch milk yields carefully during the grazing season and as soon as any sharp fall in yield or condition of cows occurs either change the grazing or increase the area of crop being grazed.

Do not expect high-yielding cows to walk long distances and be beasts of burden. Under muddy conditions do not expect close grazing of kale; let the cows have the leaf and use dry stock to clear off the stumps.

One of the biggest handicaps to fuller exploitation of grazing in its various forms is mud and discomfort; if cows have to walk through mud every day, lameness is sooner or later likely to occur—with very severe consequences on milk yields.

In very wet weather it is better to harvest, cut and cart kale for the cows or resort to silage or root feeding. Kale can be handled with modern forage harvesters and self-unloading trailers, as in zero-grazing.

Make Food Changes Gradually

Finally, always make any change in the diet of the cow as gradual as possible, particularly at the beginning and end of the grazing season.

Recent research has shown that the rumen or first stomach of the cow, in addition to having a storage function, is the site of considerable pre-digestion of food carried out by bacteria and other micro-organisms appropriate to the foods being consumed.

Change of diet requires a change in the bacterial flora—which takes time. Furthermore the physical condition of the food in the rumen is very important in promoting regular cudding; a cow can only digest what food is cudded properly.

So aim to keep the rumen in an active and healthy state by avoiding large feeds of individual foods (e.g. kale or silage) which may upset the normal "texture" of food in the rumen and so cause metabolic disturbances, of which bloat is an excellent example.

Care is particularly needed to ensure a gradual adjustment of the cow's dietary to an increasing intake of grass or forage crops when changing over from winter to summer feeding or vice-versa.

Feeding roughage or cud-promoting food is invaluable in this respect and in maintaining butterfat percentages.

CHAPTER III

ART AND SCIENCE OF MILKING

FEEDING and milking; those are the herdsman's two main jobs. The first we have already discussed. Now we can turn to the second, and deal with it not just from the actual job of taking milk from the cow, but from the wider aspect of preparing a cow for her lactation and doing everything possible during the lactation to get the maximum economic yield.

Formerly the art and science of milking began with steaming up during the dry period, which is always associated with Professor Boutflour's work. The ME system, however, lays emphasis on the need to start preparing for the next lactation more gradually and over a longer period extending forward well into the latter half of the previous lactation.

During the last four months of pregnancy the unborn calf grows rapidly and extensive growth or renewal of udder tissue occurs. This points to the need for adequate nutrition for the cow or heifer as she approaches calving, well provided for by the ME system.

STEAMING UP

It is well worth revealing the essential features of the author's steaming-up programme, even though scientific sources suggest the more gradual preparation of the cow for the next lactation.

The practice of steaming up was designed to meet these needs and to provide reserves of body tissue on which the cow can draw when milk production begins.

Steaming up is best achieved by feeding a balanced feed of concentrates during the last six weeks of pregnancy, gradually increasing the quantity fed. The level at which steaming up should be pitched should be governed by the estimated milk yield of the cow and her bodily condition when steaming up begins. The following is a basis:

THE AUTHOR'S STEAMING-UP GUIDE

Weeks before Calving	*Expected peak yield of cow (litres)* 13·6	18·2	22·7	27·3	31·8	36·4	
6	–	–	–	0·9	1·8	2·7	kg of concentrates
5	–	–	0·9	1·8	2·7	3·6	
4	–	0·9	1·8	2·7	3·6	4·5	
3	0·9	1·8	2·7	3·6	4·5	5·4	
2	1·8	2·7	3·6	4·5	5·4	6·3	
1	2·7	3·6	4·5	5·4	6·5	9·2	

Here are some special points to note in connection with the steaming-up table:

1. Cows in good condition should receive 1 kg less concentrates per day; cows in lean condition receive 1 kg extra.

2. Heifers should not be steamed up beyond a peak 3-4 kg of concentrates per day.

3. The concentrates should be "balanced" for milk production and laxative in effect. A proportion of palm kernel cake—up to 50 per cent—is recommended; beans should be fed only in small proportions, not more than 25 per cent.

4. The cows must be individually fed, allowed good quality bulk foods (hay, silage, kale, etc) to appetite; concentrates are preferably fed in the cowshed or milking parlour so that the cows become used to their surroundings and to being handled.

5. The udders should be examined daily for any possibility of mastitis developing.

6. A record of calving dates is obviously essential and should be posted up in the cowshed or parlour.

RELIEVING UDDER CONGESTION

Some cows and heifers may develop swollen and inflamed udders well in advance of calving. A heavily congested udder is to be avoided; if steaming up is at too high a level and steps are not taken to relieve this congestion, the udder becomes unduly large and pendulous.

This congestion can be relieved by milking before calving, or changing the cow's diet to a less "forcing" one—substituting a bran and beet pulp mash for the normal concentrate feed.

It is a mistake to let dry cows get too lean, or to give cows an inadequate dry period and then try to remedy the situation by heavy steaming up.

Where pre-calving milking is carried out do not begin until the teats are stiff and the sole of the udder quite firm, then ease the udder twice a day until within 12-24 hours of calving—as far as can be judged.

The milk withdrawn is true colostrum and some should be kept in readiness (preferably in a refrigerator) for feeding to the calf at birth.

I have not found it necessary to pre-milk cows for longer than three to four days before calving provided close watch is kept on the development of udder inflammation and an appropriate adjustment in feeding is made, and provided the calving date is accurately known.

Massage of the udder (particularly with heifers) will help to reduce congestion, as will exercise. But do keep steaming up to a reasonable level. The development of a highly-inflamed and, therefore, a painful condition in the udder may make for difficulties in milking and lead to bad milking habits such as paddling of the feet and fidgeting during milking.

After calving, feed on a laxative diet until udder inflammation has subsided. Be guided by the cow's appetite as to when full feeding of concentrates begins.

Should a cow fail to eat readily three to four days after calving, this may indicate post-calving inflammation of the genital organs and require veterinary assistance (see page 120).

Leaving the calf to suckle the cow for 2 or 3 days is not advised, because when the calf is removed the cow becomes very upset.

Drying-Off Technique

To ensure at least a six-week dry period should be the aim in drying off.

Cows will dry off when milking or suckling ceases as milk accumulates in the udder over a period of 48-72 hours and re-absorption of the milk occurs. This is a natural process but can present problems where disease (mastitis) is present in the udder, often in a sub-clinical form which flares up as milk accumulates in the udder.

As a result of research work at the NIRD, the recommended procedure to combat mastitis if present, or to confer protection to the dry udder for at least four weeks of the dry period (see summer mastitis, below), is as follows:

1. At the last milking, milk out clean and infuse a long-lasting type of penicillin (obtainable through your veterinary surgeon) into each quarter.

2. Dip the teats in an iodophor solution (see page 51). Then leave the udder to stock up; do not milk out and the udder will dry up within a week.

Some cows within 6-8 weeks of their next calving may be still milking at a level of say 9-13 litres per day, when abrupt cessation of milking could cause distress to both the cow and the herdsman. To assist drying off, such cows should be kept indoors and given hay only, or receive no concentrates, being milked once a day for 3-4 days and then perhaps every other day until yield falls to 1 gallon or less, when complete cessation of milking and antibiotic infusion of the udder (as mentioned on the previous page) can be carried out.

The dry period is an excellent opportunity to tackle mastitis of the chronic type caused by *streptococcus agalactiae*, but the acute *staphylococcal* type and summer mastitis can present an as yet unsolved problem during the last 2-3 weeks of the dry period.

Summer Mastitis

Under no circumstances should dry cows escape a daily inspection. Immediately any hardening of a quarter, or stiffening of a teat, or loss of appetite is observed, call in veterinary help else a valuable cow may be lost to milk production.

Further research on summer mastitis is being pursued, but in the belief that it is mainly fly-borne, keep dry cows away from low-lying, tree-studded pastures.

Treat any sore teats with particular care—especially if the sores are on the end of the teats. (See reference to wounds on page 122.)

Handle Heifers Carefully

To have cows pleasant and easy to milk requires quiet and sympathetic handling of heifers at calving time. They are often excited, nervous and easily upset.

Before calving, all heifers should be familiar with the cowshed or milking parlour and the sounds common to milking time. They should be used to having teats and udders handled and to the presence and sound of a milking machine.

When you first start to milk a freshly-calved heifer don't put on the teat-cups and then leave her unattended. Teat-cups falling off can frighten a heifer and make her kick. Be patient for two or three milkings, and the time you spend in coaxing her to a ready let-down of milk will be well rewarded.

Heifers kick mainly because of fear rather than inborn viciousness, but some nasty-tempered individuals do occur. Such temperament is often inherited, and no breeder of dairy cows ignores such evidence. Daughters of different bulls, as shown by progeny testing stations results, vary considerably in temperament; but the good stockman can do a great deal to get quiet, good-tempered cattle.

Once the heifer or cow is happy to be milked, then the machine can be relied upon to do the rest, but a vigorous let-down is fundamental.

Milking Routine

So much for initiating an animal to milking; now a few words about the routine of milking—which should be designed to ensure quiet, happy cows, clean milk production and adequate hygiene to prevent the spread of disease—principally mastitis.

The sequence of operations prior to, during, and after milking performed by the herdsman with the willing co-operation of the cows being milked should follow a fixed routine at each milking. Cows are creatures of habit and resent change.

In milking parlours the admission of the cow to the milking stall should be done as quietly as possible, so that she is in no state of fear, and in cowsheds a hurried approach should be avoided. Speak to the cow as you prepare her for milking.

The first operation is to take the foremilk from each teat into a strip cup, then wash the udder and teats with a warm water jet and wipe dry with a paper towel. Alternatively, the udder can be cleaned with an udder cloth using a bucket of warm water. Both water and udder cloth must be changed frequently, according to the degree of udder contamination.

In both cases a suitable anti-bactericidal udder wash, such

Driving cattle involves hazards when the rate of progress exceeds the natural easy pace of the slower members of the herd. A well-routined herd moves in a long line, each animal at its own pace and in its proper place.

The two popular ways of controlling grazing—paddock grazing (top) with cows run at about 25 cows/hectare clearing a paddock in 4-5 days, and strip-grazing (bottom) with cows at about 200 to the hectare clearing a daily strip.

Calf housing can be purpose-built, as in the top picture with calves housed in individual pens. But old buildings can be suitably converted, as in the lower picture where a straw-bale false ceiling has been used to provide comfortable pens for group-housed calves.

Calf rearing costs can be cut by rearing calves at grass using a paddock grazing system.

The cubicle system of housing provides cows with the privacy and comfort they require and allows them to keep themselves reasonably clean. These diagonally-sited cubicles make maximum use of the width of building available.

Daily scraping of the passages in a cubicle house can be eliminated by the use of slatted passageways.

An alternative to slatted passageways is a metal grid floor such as the one here in the NIRD cubicle house.

One of the cheapest forms of cow accommodation is the cow kennel. Overhanging roofs of two rows of kennels keep out the rain but allow adequate ventilation in this kennel set-up.

This kennel-housed herd has grass cut and carted to it for self-feeding direct from the trailer—see right of picture.

One of the problems of no-straw cow housing systems is slurry handling. One answer is the slurry tanker (top picture). Another but rather more expensive solution is to pipe the slurry direct on to the land through an irrigation system. (bottom picture).

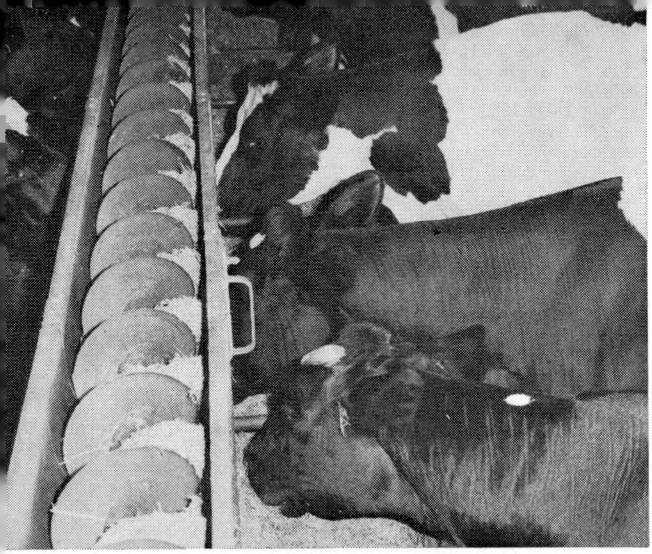

Self-feeding and mechanical feeding systems have eliminated a lot of the labour involved in the feeding of silage. Photos show: top, an auger system which reduces silage feeding to a push-button operation; centre, the help-them-selves approach with self-feeding controlled by a tombstone barrier; bottom, the mechanisation of conventional silage feeding by means of a self-unloading trailer.

Alternative bulk feeding system is one based on straw in conjunction with balancer concentrate and barley.

Urea-based balancers are also used with straw feeding systems, provided on a self-feed basis either in block form (above) or in liquid form through a special feeder (right).

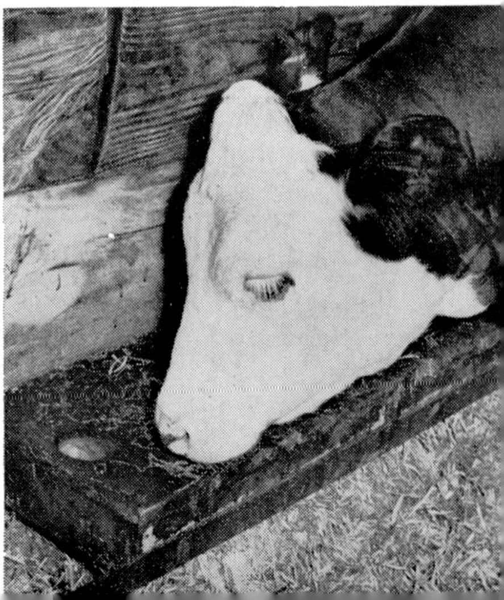

as hypochlorite, quaternary ammonium compound or iodophor, should be added to the washing water according to makers' instructions.

If evidence of mastitis is found by clots in the foremilk, or pus which discolours the milk, the cow should be marked (e.g. with a coloured aerosol spray) for subsequent treatment. Marking will also ensure that her milk is not added to the bulk milk at that milking or for the four following milkings if she receives antibiotic treatment, as is required by the antibiotic-residues-in-milk clause incorporated in MMB contracts since April 1966.

Washing and handling of the teats stimulates the "let down" of milk, due to the release of a hormone into the blood stream from the pituitary gland which causes contraction of muscular tissue in the udder. This acts rather like the squeezing of a sponge, raising the pressure of milk in the milk cistern above each teat to assist in the ejection of milk through the teats under the action of the vacuum applied to the end of each teat when the teat-cups are attached.

A cow not properly stimulated, or frightened, will not give a full "let down" of milk.

Attach the teat-cups by holding the cluster in one hand and using the other to lift each teat-cup in turn into the vertical position. Once in that position, suction will be restored to it and it will slip readily over the teat, particularly if you hold the teat-cup near the top and use your first finger to direct the teat into it. The teat-cups should be applied within two minutes of washing being completed.

This is most important as some cows have a very short period of "let down". If teat-cup application is too long delayed milk yield at that milking will be much reduced, and if this practice continues, such a cow will soon dry up and show a very short lactation.

Any cow which, for no apparent reason, shows a decline in yield, particularly in early lactation when she ought to be rising in yield, should receive particular attention at milking. She may be one of those cows with a very short "let down" period —the teat-cups should be applied to such cows immediately after udder washing.

Apply the teat-cups to the back teats first. The teat-cup

cluster can be supported by a back cord if necessary.

Whilst the milking machine is operating on one or more cows the next cow or cows can be prepared, but a close watch should be kept on the sight glass of each machine and as soon as air bubbles appear—indicating a slackening of the milk flow—stripping should be commenced.

This is performed by pressing gently on the teat-cup cluster claw with one hand and massaging each quarter (particularly the hind quarters) between the thumb and the fingers of the other hand, just above each teat.

As the milk cistern above each teat empties, the "back flow" of milk between fingers and thumb will lessen and disappear. The quarter is then milked out.

Remove the teat-cups immediately stripping is completed, closing the vacuum tap first in order to prevent dirt being sucked into the open ends of the teat-cups on removal.

It should be also remembered that clean cows save time spent washing, but udder stimulation by massage is still recommended with clean cows to give a full "let down" of milk. One advantage of 'cubicle' housed cows is that they keep remarkably clean (see page 69).

Organisation of Machine Milking

A summary of the units/man and throughput of cows/hour expected for different milking systems is shown in the following table—realising that the calculated throughputs are rarely achieved in practice.

Naturally slow-milking cows exist in every herd which reduce throughput in cows/hour or increase the number of units required. And with newly-calved cows the work routine time tends to be increased by the extra attention required.

System of Milking	*Units per man*	*Cows per hour*
Bucket plant—milk carried to dairy or tipped into churns in shed, concentrates fed at milking	2	16
Bucket plant—milk carried to dairy by pipeline, no concentrates fed at milking	3	24
Parlour milking—abreast or tandem	3	24
Parlour milking—herringbone	4	36
Rotary parlour milking	5-9	80-150

In parlour plants where feeding of concentrates occurs at milking, a 2 stall/1 unit lay-out is desirable in order to give each cow sufficient time in which to consume her concentrate ration. Cows will consume about 2 kg of cubes or 1·5 kg of meal in 5 minutes—a 2 stall/1 unit layout gives 10-12 minutes feeding time compared to 5-6 minutes in the 1 stall/1 unit layout.

It is obviously difficult, therefore, to feed cows giving in excess of 27 litres a day sufficient concentrates without allowing a longer feeding time in the parlour, which can upset the parlour routine.

Such cows are best yarded separately and milked last, or given separate feeding outside of the parlour. This is a strong argument for running the cows in two groups, formed at calving time as suggested for cowshed housing (see page 30), but is not easy under self-feeding systems unless two feeding faces are available at the silo.

The temptation to re-shuffle groups during lactation should be resisted because yield is often adversely affected as a result of movement from one group to another.

Hints on Using Milking Machines

Should a cow be "blind" in one quarter, the teat-cup concerned is doubled back so that all suction is cut off and kept in that position with a rubber ring.

Most milking machine cups today can be fitted with a rubber ring at the top of each cup to prevent "clanking" of the cups as they are taken off. This avoids any unnecessary noise and kicking on the part of the nervous individuals, particularly heifers.

In frosty weather any cracked or chafed teats should be treated with a non-greasy udder cream before the cows are turned out again, and all teats should be wiped free of any adhering milk at all times with clean damp towels, frequently changed.

As a means of mastitis control, dip the teats in an iodophor teat dip or spray with an aerosol containing dichloropen with iodophor in an alcohol base. This, with the use of a long-lasting penicillin infusion of the udder at drying off, has given significant control of mastitis, initiated by work done at the NIRD.

In some herds mastitis is a recurring problem through injury to the udder tissues arising from faulty use or maintenance of milking machines. Such injuries can be caused by:

1. Too high a vacuum, *or* a fluctuating vacuum resulting from insufficient pump capacity (see page 85).

2. The use of hard-topped teat-cup liners. I prefer a soft-topped stretched liner to the moulded hard-topped type of liner.

3. Leaving the teat-cups on after milking is completed (over-milking), as mentioned earlier in this chapter—most cows milk out in 4-5 minutes but slow milkers may take up to 10 minutes. This results in the teats being sucked too far into the liners and may indeed cause damage to the ends of the teat canal. Once the milk pipe is empty of milk, the teats are subject to the full vacuum effect without the "cushioning" effect of milk in the pipeline or the teat canal.

Before attempting to restrain a cow or heifer from kicking, try to discover why she kicks.

Feeding at milking time is often used to make the cows happy and at ease. But this means an additional task superimposed on the actual job of milking, and if it can be done before milking so much the better. Let-down of milk is a matter of training, and feeding is not necessary for the purpose —although in many milking parlours with outlying herds it is unavoidable.

Preventing a Cow from Kicking

As a last resort, when patience is exhausted, the cow can be prevented from kicking by:

1. Tying a rope in figure-of-eight fashion around her hind legs above the hocks. This is very effective, but it may—until the cow gets used to it—involve some "hold up" of milk.

2. Fasten a rope around the cow's middle, forward of the udder and across the small of her back. This is a less drastic method than 1.

3. Hold up the cow's tail in a well-nigh vertical position, and at the same time try to soothe her by scratching her rump. This is the least drastic method and is often quite effective with machine milking.

4. Halter the cow and tie her head up high. This method is helpful when a herdsman is working single-handed.

These methods may also be used in restraining a nurse cow when suckling calves.

The rate at which a cow releases her milk is a matter of heredity and training. Naturally "slow" milkers exist with constricted teat orifices; but naturally "fast" milkers can, if subject to a poor milking technique, become desultory in their let-down of milk.

Leaving teat-cups on too long not only encourages slow milking but makes the delicate membranes lining the teat canal and the milk cistern subject to the full influence of the milking machine vacuum. They are then susceptible to injury and mastitis infection is much more likely as a consequence.

Good milking is quick, quiet and thorough. No machine can do its job without the full co-operation of the cow and it is the herdsman's job to obtain that co-operation by gentle persuasion.

Only when cows are happy to be milked will they give of their best. Milk records—which we will now go on to discuss—reflect the milking skill of the herdsman, the milking ability of the cows and the feeding programme.

Milk Recording

Milk recording has a most important part to play in good herdsmanship. It is usually a herdsman's responsibility, but it should in any case be used by him as a guide to proper herd and cow management.

The objects of milk recording are:

1. To provide a guide whereby accurate feeding according to yield can be carried out.

2. To give a check on the daily health and breeding regularity of the cow as well as on managerial factors such as feeding and milking routine.

3. To provide the means (with butterfat testing) of assessing the productive ability of breeding stock and give the information needed in breeding better dairy cattle.

4. To give production histories essential in the sales of pedigree stock, particularly bulls.

Under good management, where feeding, milking and general welfare of the cattle are satisfactory, a study of the milk records provides a background against which mistakes in management can be detected and adjustments effected.

When the recorded milk yields of cows are plotted on a graph week by week, a composite lactation curve can be drawn which typifies the milking performance of the average cow in the herd. Such a curve, drawn from my own data, indicates the following general features:

1. After calving, a cow's milk yield will rise steadily for a period of from four to six weeks when the yield reaches its peak.

Failure to maintain this peak yield is often due, not only to inadequate feeding, but to an overstocked udder when too long an interval is imposed on the cow between milkings.

2. The daily peak yield of heifers is approximately 1/220th of the 305-day lactation yield, 1/200th in the case of older cows.

Thus a heifer giving 18 litres a day at her peak should produce a lactation yield of 4000 litres, and a cow giving 27 litres a day at her peak should have a lactation yield of 5400 litres, on the average.

3. Once the peak yield has been reached the subsequent decline in yield with cows is at the rate of approximately $2\frac{1}{2}$ per cent per week or 10 per cent per month.

Pinpointing the short-lactation cow

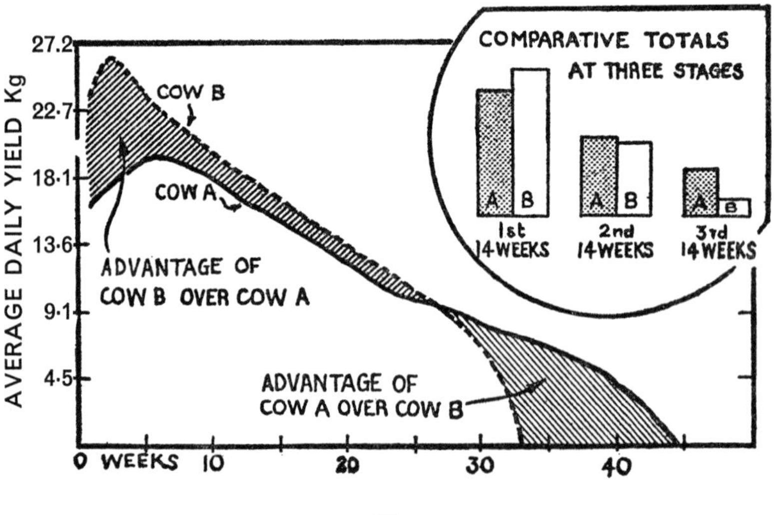

Heifer yields decline more slowly by only 7-8 per cent per month.

A more rapid decline in the yield of the herd as a whole would indicate under-feeding or poor milking methods.

In early lactation the cow will draw on her own body reserves to remedy an inadequate diet. Later, the milk yield will show a rapid decline in proportion to the inadequacy of food available. Generally, where concentrates are rationed to yield, this indicates home-grown foods of poor quality.

Once management conditions are established graphically by means of lactation curves, the naturally short-lactation cow can be positively identified with the certain knowledge that no management factor (e.g. under-feeding) is responsible.

The milk records of two cows under the same management are shown in the diagram (page 54). The two curves illustrate the naturally short-lactation cow (B) compared with a more persistent milker in cow (A). The latter is by far the more desirable type, being easier to feed on bulky foods owing to her lower peak yield and greater persistency.

Intelligent use of milk records is thus not only an index of management generally but is the basis for culling low-yielding and short-lactation cows. Cows that have flat lactation curves possess a very valuable characteristic—milking persistency.

Recording Routine

Now a few details about the routine of recording.

1. How to join. Write to the Milk Marketing Board, Thames Ditton, Surrey, or to the Scottish Milk Marketing Board, Paisley PA3 1TJ.

2. Milk-recording Year. Runs from mid-day on Sunday, at or about 1st October, for the following 52 weeks.

3. Qualifying Records. A qualifying record for an animal's inclusion in the herd average is made when the cow or heifer calves in the herd and is milked for at least 200 and not more than 305 days from the fourth day after calving. Also the previous calving date must be authenticated.

4. Calving Index. Provided the previous calving date of an animal is known, the interval between the two calvings im-

mediately preceding the lactation, is known as the calving index. An average for all cows in the herd becomes the herd calving index, which should approach 365-385 days. If it exceeds 400 days, there could be a fertility problem or, alternatively, some feature of management might need adjustment.

National Milk Records

The recording scheme operating under the heading of National Milk Records (NMR) in England and Wales is the Monthly Recording by Statement system (MRS), and this has now superseded all previous NMR schemes. It is a nationally and internationally recognised scheme, supplying authentic records for such things as bull proving, pedigree records, sales catalogues and breed society registration.

The attraction of the computerised recording scheme from the member's viewpoint is its speed in the presentation of updated information. A contract recorder visits each producer once a month, at which time the evening and morning milk from each cow is weighed and samples taken for butterfat and protein testing. Service and calving dates, purchases and disposals are also noted. The producer will then receive within about 15 days a computer statement which presents the up-to-date herd position at a glance. Complete 305-day standard lactation certificates are issued automatically.

In view of this rapid turn-round time, the computer system has the potential for developing the managerial aids necessary to efficient herd recording, and to meet this need the statement layout is continuously under review to incorporate new features, A separate action sheet is supplied each month, listing cows to breed, cows to calve, cows to dry off and cows to pregnancy check.

A recent (1977) innovation in milk recording has been the introduction of the Production Index. This places each cow in the individual herd on a percentage scale in which 100 represents the herd average at the time. The index is based on each cow's production records, and is regularly updated. It can be helpful when making decisions on culling.

The monthly fees for NMR from October, 1977 are as follows:

Herd size (cows)	*MRS (including fat and protein testing)*
	£
1– 25	8.45
26– 35	9.80
36– 45	11.50
46– 55	12.85
56– 65	14.45
66– 80	15.70
81–100	17.90
101–125	20.95
126–175	23.85
176–225	26.75
226–275	29.80
276–350	32.70
Over350	35.60

Additional contractor: Per visit £8.04. *3 x milking:* Per visit £4.25. *Divided herd:* annual fee £4.05.

Farm Management Scheme

Farm Management Recording (FMR), which was introduced in April 1972, superseded the old Private Milk Records. The purpose of the system is to provide the member with information vital to the control and management of his dairy herd, and the figures and records produced are for his own use and are not eligible for breed society registration or sales catalogues.

The basic scheme provides the member with recording stationery for either daily, weekly or monthly recording, together with annual herd summaries which list each cow's performance. Health cards and individual cow record cards listing service and calving details are also provided. In addition to these features, the scheme provides the member with stationery on which, each month, he may work out important management efficiency factors concerning grain and concentrate usage in relation to milk production and financial return. The basic scheme provides for a fieldsman to visit the member's farm four times per year to carry out the calculations necessary to update his milk records. The fieldsman also brings up to date the herd summaries and individual cow record cards from the information recorded by the member.

As well as this basic service the member can select the following optional extras; computerised updating of weighings taken monthly by the member; analysis of milk samples, taken by the member, for butterfat percentage; and analysis for butterfat and protein in samples taken each month by the member.

The monthly fees for FMR from October, 1977 are as follows:

Herd size (cows)	*Basic service non-computerised)* (A) £	*Basic computer service* (A+B) £	*Optional fat testing* (C) £	*Optional fat and protein testing* (D) £
1– 20	1.25	2.15	1.45	1.90
21– 30	1.25	2.15	1.80	2.45
31– 40	1.55	3.50	2.05	3.15
41– 50	1.55	3.50	2.60	3.75
51– 60	1.55	3.50	2.95	4.35
61– 75	1.90	5.55	3.55	5.30
76–100	1.90	5.55	4.55	6.90
101–150	2.40	7.85	5.55	8.95
Over 150	3.15	10.55	7.45	11.65

Options available: A, A+B, A+C, A+D, A+B+C, A+B+.D

INTERPRETING RECORDS

From large numbers of records it is possible to determine several factors that act as a guide to interpreting records clearly and are an aid in improving management. From an original study by Sanders we learn the following:

Age at Calving

Milk yields generally rise with age up to the maximum at the fifth or sixth lactation and then decline with increasing age. In classifying yields as an indication of merit, breed societies take this fact into account.

Previous Dry Period

A very short dry period tends to depress the following lactation yield. The optimum dry period would appear to be about 50-60 days.

A longer dry period may give a higher subsequent lactation yield but not sufficiently large to offset the loss of milk from the viewpoint of annual production.

Service Period

To calve regularly at twelve-monthly intervals, a cow needs to be in-calf by the 12th to 14th week after calving.

To delay service beyond this period means a higher lactation yield but a wider calving interval, whereas earlier service means a lower lactation yield and a calving interval of less than a year.

If regular calving of each cow at the same time of year can be achieved, the management of the herd is likely to be much more efficient.

Season of Calving

Most dairy farmers would hold the opinion that autumn-calving cows give more milk than spring or summer-calvers. Sanders found 11 per cent difference in favour of October calving compared with June calving.

Better grass management during the grazing season, and the use of high-quality home-grown foods in winter feeding might significantly reduce this difference.

Frequency of Milking

Twice-daily milking is most common, but an increase of production can be expected where—

1. Milking intervals are evenly spaced (i.e. 12-hourly intervals instead of, say, 14 hours and 10 hours), particularly with cows at the height of their production.

2. By further removing the depressing effect of udder pressure on yield, three-times milking can be expected to raise production above twice-daily milking by 20 to 25 per cent. This applies especially to high-yielders and potentially high-yielders, provided that no milking interval exceeds nine hours.

Thrice daily milking involves more labour but can pay with high yielders and in winter months when milk prices are high. However, there is the possibility with twice-daily milking of making the night and day intervals more uniform by milking high yielders first in the morning and last in the afternoon.

Lactation can be extended, yields increased and udder shape preserved by either thrice-milking or equalising the intervals between twice-milking.

Milking Routine

It is very harmful to make any change in the milking routine after the twelfth week of lactation—from hand to machine milking, from thrice- to twice-daily milking or by transferring the cow from one group to another.

In late lactation when yield has dropped the milk loss is relatively small. In very early lactation the daily yield suffers temporarily but recovers, but after the twelfth week recovery is poor. The best time to effect fundamental changes is the dry period.

Facts that Affect Records

In interpreting official records all these factors are taken into account, and a common basis for comparison between individual cows and between herds can be determined mathematically by appropriate correction factors. But, for practical breeding work, the important matters to bear in mind when assessing the value of a record are these: correction for age, service period and frequency of milking.

There are, however, other influences—not capable of definition in an official record—which must be borne in mind.

The first is the level of concentrate feeding and the quality of the home-grown bulk food consumed—in other words the general plane of nutrition at which the animal has been fed. Today this is, to a very large extent, a reflection of the general fertility level of the farm and the skill with which the land is being farmed—as reflected in the quality of home-grown foods.

Records obtained under low fertility—e.g. hill farming—conditions, are not generally so high as those on the more fertile vale farms, even though the cows in both cases may be of equal milking capacity.

Secondly, the skill of the herdsman in handling cows, the degree of steaming up, the standard of milking technique used, and freedom from disease can make a most significant difference to milk yields.

Thirdly, the system of housing employed, or the method of wintering cows will affect yields. Generally, under extensive management (for example, milking by bail or in a parlour) large herds will record lower yields per cow than small herds given greater individual attention.

Therefore, to interpret a milk record intelligently when buying cows, or in determining breeding policies, calls for judgment as to how far the record is a fair indication of the cow's inherent productive ability.

Under the official scheme space is provided on the milk-recording sheets for noting dates of calving, dates when cows go dry, services, and bouts of sickness, disease attacks, and so on, and such notes should be conscientiously made. They assist in resolving herd problems.

When they are transferred to the MRS computer system a complete history of each cow over each lactation is built up.

This is most valuable information to the owner and to his veterinary surgeon. From such information a calving schedule can be drawn up, cows can be ensured an adequate dry period, and those cows which repeatedly return to service, or calve prematurely can be readily noted and appropriate veterinary assistance called in.

Some herdsmen keep a record of services in a separate book. Others prefer visual charts which automatically record failure to hold, reservices and calculated dates of calving, and also date on which steaming up should commence. Both methods are obviously applicable to beef herds and again should be most conscientiously compiled.

A Check on Food Consumption

In addition to milk recording every herdsman should appreciate the importance of keeping a check on food consumption, particularly of expensive concentrated foods. In milk-selling herds, the quantity of concentrates fed is often far in excess of what is supposed to be fed, with consequent loss to the owner in profitability.

To feed concentrates accurately is the herdsman's responsibility.

When rations are mixed, the quantity for a definite period should be made up in relation to stock requirements over that period, and a real effort made to relate consumption to production.

This is the kind of statement that can be drawn up:

Example of food control in a herd of 30 *cows in milk*, 6 *dry cows*

Sales of milk for previous week (wholesale)	2,700 litres
Add milk fed to calves and supplied retail	190 litres
Total produced	2,890 litres

Subtract from 2,890 litres the number of litres of milk produced from home-grown foods fed at M + 4·5 litres level—

Total produced	2,890 litres
30 cows in milk at 4·5 litres/day	955 litres
Therefore milk from concentrates equals	1,935 litres
Concentrates required at 2·5 litres per kg equals (approx) ...	774 kg
Add concentrates for "steaming up" 6 cows at 3·6 kg per cow per day	150 kg
Total consumption for week	924 kg

The concentrate allowance for each individual cow is determined by her milk yield and should be chalked up near her stall or marked on her individual feed bucket according to the method of rationing used.

When the consumption of concentrates is controlled in this way milk yield from the herd as a whole is influenced by how adequate the home-grown foods are in meeting the assumed level of maintenance and the first gallon.

When Yields Fall Too Quickly

Too rapid a decline in milk production could arise from any of the following causes:

1. Adverse weather conditions, particularly much wet and windy weather.

This causes an increase in maintenance requirements, as would long walks to graze off kale in muddy conditions.

Dry, cold weather is of much less consequence in its effect on milk yields in that the cow can withstand quite low temperatures, provided she obtains sufficient water. But out-wintered herds should have dry-lying land and shelter from wind.

2. Inadequate food being eaten.

This may be due to waste in feeding by muddy conditions underfoot if cows receive any part of their feed outdoors or, in frosty weather, failure to eat sufficient kale or roots. Silage is a very useful substitute for green crops or roots in these circumstances.

3. Too little food being allowed.

In periods of adverse weather, maintenance rations should be increased by anything up to 20 per cent unless the cows are fed entirely indoors.

4. Home-grown foods of low feeding value or of low palatability.

Silage of a high moisture content and mouldy hay are unpalatable; accurate feeding of such foods is impossible unless some idea is gained of what the cows are actually eating.

Again low-quality hay and silage of poor feeding value may be little more than a belly-filler. If such is the case the remedy is to "boost" the feeding value of these home-grown feeds by including, say, 1 kg of oats or beet pulp as a hay substitute rather than by increasing the quantity of low-quality foods.

Once a cow is past her peak yield, normal decline in milk yield—as we have previously seen—is about 2½ per cent per week. On a 20 litre yield this is 0·5 kg per day per week.

Low Butterfat

In England and Wales milk must have at least 3 per cent butterfat (in Scotland 3.5 per cent) and there is now a price incentive encouraging producers to pay attention to the fat percentage of the milk.

Producers who are in difficulties should be aware that some strains of cow are poor fat producers as well as some breeds being poorer than others. Old cows tend to produce low fat milk and a regular intake of heifers into a herd keeps up bulk fat tests.

As lactation progresses and yield rises fat percentage falls and, therefore, if most of the cows in the herd are at peak yield the bulk milk is likely to register lower fat percentage. The strippings contain a higher fat percentage than the milk extracted first and therefore stripping properly is important.

A milk-equaliser fitted below the cooler distributes all the milk between several churns and this reduces the risk of one churn's milk being filled only with low-fat milk.

Cows which are maintained in good condition and which calve in good order will give milk of better quality than lean cows. And certain foods, like palm kernel cake or coconut cake, tend slightly to raise the percentage whereas cod-liver oil if fed over 60 g per day will depress it.

When cows are on lush grass in the spring, the fat percentage tends to fall. This can be mitigated by feeding 2 kg hay or straw before going out to grass or by providing more mature grazing.

Uneven intervals between milkings are responsible for some loss of fat yield. Fat percentage is usually lower after the long period between milkings.

The solution to a low-fat problem lies in identifying which of the above-mentioned factors are operating and making the necessary adjustments.

Low Solids-Not-Fat

Progressive underfeeding is mostly responsible for low solids problems in dairy herds—in late winter, typically. In a dry summer, as a result of grass shortage, the problem may reappear and, again, as a result of starting to feed concentrates too late as a supplement to autumn grazing.

Sub-clinical mastitis infection of a herd can be responsible for a reduction in snf of as much as 0.2 per cent.

Old cows and cows at peak of lactation tend to produce milk of low snf and some strains of cattle are characterised by low snf per cent so, as is the case with fat percentage, it is a matter that can be taken care of to some extent by progressive breeding and recording practices.

Use of Official Milk Records in Breeding

So far I have discussed the application of milk records and feeding records to cows only. There is, however, an additional application of milk records, and herdsmen who are concerned with breeding policies should know of it and make use of it.

The Production Division of National Milk Records collates all official milk records for progeny-testing purposes, and today pedigrees can be compiled showing not only milk records of female ancestors, but records of a bull's daughters which give a complete picture of performance.

Such pedigrees are illustrated on the following page.

The usefulness of a pedigree in evaluating the probable performance of an animal is thus greatly increased; but it must be remembered that at best a pedigree is a promise rather than a proof of performance, whereas milk records for females provide the proof not only of their own merit but of the breeding value of their parents.

Without milk recording, progress in breeding through progeny testing is impossible. It supplies the science which, when allied to the breeder's art in feeding, management and selection, makes real progress possible.

Beef Recording

The recording of liveweight gains in beef animals is a new and welcome development. It gives beef feeders information as to the progress cattle are making on specific feeding (i.e. on food conversion efficiency), or to their inheritance of growth capacity as in performance testing of bulls (see page 143).

Beef recording achieved national status, under the Beef

(please turn to page 66)

Pedigree of TERLING TEMPEST R.M.
Breed: Friesian Born: 23.10.42
H.B. Ref. No. 78741

	Lb	Days	%BF.
74 d. with 1st c. av.	9,584	300	3.51
55 d. " 2nd "	9,633	284	3.54
36 d. " 3rd "	11,366	287	3.55

Daughters recorded in 32 herds
Bred by: Lord Rayleigh (Farms) Inc.,
The Bury, Hatfield Peverel,
Chelmsford, Essex.

TERLING TEMPEST

SIRE Terling Olympia P.I. R.M.
H.B. Ref. No. 56303 Born: 10.1.39

	Lb	Days	%BF.
72 d. with 1st c. av.	9,378	303	3.57
59 d. " 2nd "	11,686	298	3.57
46 d. " 3rd "	13,010	294	3.55

DAM Lavenham Grey 88th R.M. R.M.L.
H.B. Ref. No. 241362 Born: 17.11.38
Ear No. 0 80347
1. 13,269 - 365 - 3.79 - 7
2. 23.10.42 - 15,062 - 327 - 3.44 - 6
3. 29. 9.43 - 14,235 - 365 - 3.61 - 8
4. 14. 2.45 - 15,293 - 365 - 3.60 - 8
5. 10. 7.46 - 13,521 - 305 - 3.15 - 5
6. 9.10.47 - 13,673 - 305 - 3.49 - 8
7. 1.12.48 - 13,321 - 305 - 3.47 - 8
8. 29.12.49 - 14,394 - 305 - 3.23 - 8
9. 19. 2.51 - 8,249 - 282 - 3.61 - 7
10. 23. 2.52 - 11,233 - 305 - 3.45 - 9
11. 7. 6.53 - 11,429 - 305 - 3.27 - 9
12. 9. 6.54 - 3,378 - 66 - 3.42 - 2
Left for non-recorded herd
18.8.54

SIRE Terling (Imp. '36) Yme R.M.
H.B. Ref. No. 47199 Born: 30.3.35

	Lb	Days	%BF.
32 d. with 1st c. av.	9,212	335	3.72
46 d. " 2nd "	9,702	335	5.65
46 d. " 3rd "	11,108	335	3.56

DAM Terling Collona 19th P.I. R.M.P.
H.B. Ref. No. 183804 Born: 17.9.33
1. 22.12.36 - 11,544 - 290 - 3.11 - 7
2. 24.12.37 - 18,284 - 328 - 3.42 - 7
3. 10. 1.39 - 16,420 - 323 - 3.40 - 7
4. 1.12.39 - 6,595 - 250 - 3.89 (Slip)
5. 14. 1.41 - 12,172 - 352 - 3.44 - 7

SIRE Terling Marmion P.I. R.M.P.
H.B. Ref. No. 44977 Born: 9.10.34

	Lb	Days	%BF.
93 d. with 1st c. av.	10,109	323	3.41
90 d. " 2nd "	12,558	314	3.34
88 d. " 3rd "	13,461	305	3.31

DAM Lavenham Grey 51st R.M.
H.B. Ref. No. 180180 Born: 16.12.33
1. 26. 7.36 - 9,813 - 355 - 3.65 - 8
2. 30. 9.37 - 13,516 - 329 - 3.34 - 6
3. 17.11.38 - 14,746 - 335 - 3.38 - 7
4. 11.12.39 - 13,481 - 317 - 3.49 - 7
5. 19.12.40 - 12,004 - 293 - 3.45 - 6
6. 2.12.41 - 14,026 - 365 - 3.72 - 8
7. 6. 3.43 - 11,360 - 365 - 3.48 - 6
Sold 18.8.44

SIRE Donald
H.B. Ref. No. 21284 F.R.S.

DAM Ymkje 27th
H.B. Ref. No. 89295 F.R.S.
10,010 - 407 - 4.23
6,710 - 326 - 4.48

SIRE Terling (Imp. '22) Marthus R.M.
H.B. Ref. No. 21533 Born: 18.5.21

DAM Terling Collona 3rd P.I. R.M.
H.B. Ref. No. 110662 Born: 5.1.26
1. 2. 3.29 - 9,589 - 356 - 3.37 - 6
2. 8. 5.30 - 14,208 - 304 - 3.08 - 7
3. 15. 5.31 - 19,299 - 362 - 3.43 - 8
4. 6. 7.32 - 20,584 - 365 - 3.77 - 7
5. 17. 9.33 - 14,684 - 324 - 3.10 - 5
6. 9.10.34 - 20,449 - 351 - 3.21 - 8

SIRE Terling (Imp. '22) Marthus R.M.
H.B. Ref. No. 21533 Born: 18.5.21

DAM Terling Collona 3rd P.I. R.M.
H.B. Ref. No. 110662 Born: 5.1.26
Records as above

SIRE Terling Favourite R.M.
H.B. Ref. No. 38201 Born: 20.4.30

DAM Lavenham Grey 36th P.M.P. R.M. R.M.L.
H.B. Ref. No. 138708 Born: 24.12.29
1. 27. 9.32 - 8,047 - 362 - 3.70 - 7
2. 16.12.33 - 10,304 - 319 - 3.42 - 7
3. 20.11.34 - 14,184 - 365 - 3.36 - 8
4. 31. 1.36 - 13,019 - 271 - 4.12 - 6
5. 30.12.36 - 14,489 - 299 - 4.04 - 6
6. 30.10.37 - 8,602 - 292 - 3.73 - 6
7. 7.10.38 - 14,344 - 293 - 3.85 - 6
8. 13. 9.39 - 6,153 - 329 - 4.05 - 6
9. 14.10.40 - 17,375 - 365 - 3.52 - 8
10. 4. 6.42 - 7,496 - 293 - 3.91 - 5
11. 7. 8.43 - 527 - 60 - 3.37 - 1
12. 4.11.46 - Kept for breeding purposes only

1 lb=0.4536 kg

Recording Association, and it is likely to be of increasing importance in the future, when gain-tested bulls may well be the order of the day, as is already the case in America. The weighbridge is as vital to the beef producer as the weighing scales to the efficient milk producer.

Interpretation of Beef Recording Data

The accuracy of liveweight data as an indication of liveweight gains is affected markedly by the "gut contents"—cattle "lose" weight by urination and defaecation during transport from say farm to market or abattoir.

In feeding trials, change of diet, by altering the gut contents, will also give changes in liveweight which are illusory as an index of liveweight gain as flesh. Succulent diets, because of the high water content, will show falsely high liveweights compared to dry-food diets, and correspondingly lower killing-out percentages or carcase weight/liveweight.

In weighing on the farm, try to reduce error by weighing first thing in the morning when gut contents are minimal, and at the same time each weighing day.

Finally, food conversion figures in terms of kg food per kg liveweight gain are open to very severe criticism unless the nature of the food is known. It would be far more scientific to state food conversion figures as estimated MJ of ME per kg of liveweight gain.

Otherwise the whole concept of food conversion ratios—an admirable index of merit in beef cattle—could fall into disrepute.

What is significant, however, from existing data is the economy in food associated with rapid growth and the wide variation between individuals. Some cattle will put on 1 kg/day liveweight gain, others only 0·7 kg, on the same level of feeding. The former would save at least six months' feeding in reaching 450 kg liveweight.

It is in the identification of high gain cattle, as with higher milk yields, that recording can play its full part as a tool in breed improvement, though, of course, the cost of achieving the gain is a highly relevant consideration.

Emphasis is now placed on the cost of 1 kg gain in liveweight and target costs are proposed, say 13p per kg in the winter months when housed and on conserved feeds and concentrates and 7p per kg in the summer months when the animals are at pasture.

OTHER MMB PRODUCTION SERVICES

(a) *Low Cost Production Services* (LCP) is the farm costings and consultancy branch of the MMB's breeding and production organisation. It involves comprehensive recording, the processing of the data, the monitoring of the farm business and the production of a summary of accounts. Consultants help to improve milk yield and grassland utilisation. They also advise regarding feed recommendations, cash flow projections and budgetting. There is a choice of services available to the farmer: Total Farm Business, Farm State involving the production of a trading account, balance sheet and annual cash flow from minimal records, Dairy Management Scheme and Herd Management Control.

(b) *Mastitis: Cell Count and Control Services* which are very useful in early identification and control of this disease.

(c) *Brucellosis:* Milk Ring Tests on the milk are reported on a periodic basis.

(d) *Pregnancy Testing Service:* Milk samples are taken 24 days after insemination and examined. Results are returned to the farmer within 5 days of receipt.

(e) *Milking Machine Testing:* Examination is made on an annual basis and modification effected and replacement parts are supplied at normal retail prices.

CHAPTER IV

HOUSING THE HERD AND MILKING FACILITIES

THE size of dairy herds continues to increase and, at the same time, whether they are the causes or the effects, changes take place in the accommodation of the herd and the handling of the cows, the milk and the manure. Reference is made elsewhere to the dramatic change in the average size of dairy herd in England and Wales during recent years and it still continues to increase at the present time.

In general, increasing the number of cows in a herd emphasises the need for housing the herd in winter in an economic and efficient manner. Large herds poach good grassland in wet weather already referred to on page 37 and create muddy roadways where a small number of cows would create much lesser problems. Consequently, housing considerations are very important.

Accommodation for the cows must be adequate for their comfort, economical in straw-use and convenient to operate with regard to feed provision and manure disposal and all these requirements must be met at an economical cost. Inevitably, invention and innovation have made their contribution in the form of slatted floors, cubicles and kennels. The solution of some old problems has created new problems. Example, slurry-disposal.

Slatted Floors

It has been found possible to increase greatly the stocking

density of buildings and to eliminate the need for bedding material by replacing the traditional floor by slats—concrete, iron or hardwood—spaced with gaps between them so that the manure drops through and is trodden through these gaps into a storage-pit below. The pit is emptied as often as is found to be necessary.

This idea was being adopted fairly widely, though there were problems with regard to ventilation, cleanliness and comfort for the cows, when another novelty—cubicles—caught the imagination of dairy farmers. In the last few years cubicles have been widely adopted and are found to provide reasonable comfort cheaply and economically. The cows keep themselves reasonably clean.

Cubicles

The cows are loose-housed in covered accommodation and they have free access to adjacent rows of resting pens, cubicles, each of a size suitable for one cow only and so constructed that the cow cannot turn around and foul the area.

Basically, simple partitions, 1·7-2·3 m long and 1·1-1·2 m high, are placed 1·2 m apart either at right angles, or obliquely, to the walls or alternatively head to head in a block centrally in the covered area. Typical dimensions by breed of cow are:

Breed	*Length*	*Breadth*
Friesian	2·3m	1·2m
Ayrshire and Guernsey	2·1m	1·1m
Jersey	2·0m	1·1m

It is advantageous to have movable head rails, heel bars and brisket boards so that initially the cows have extra room while they get used to using them. Later these are adjusted so that no dung falls in the lying area. The passage between rows of cubicles should have a minimum width of 2·1 m and slatted flooring of this passage eliminates the need for daily scraping. The cubicles only are bedded with straw or sawdust.

Cow Kennels

These provide a very low-cost form of cow accommodation, comprising covered rows of cubicles placed within an open concrete yard. In the southern parts of England and in sheltered situations these very simple units promise to resolve the problem of winter poaching of the grazing areas where herd size has been increased. The slurry disposal problem remains and the use of tankers or irrigation equipment is often resorted to.

Milking Machines

There are five main types of plant:

1. The bucket-type of plant with a tapped overhead vacuum line installed in the cowshed.
2. The mobile milking plant used in cowsheds and delivering the milk direct to churn (or bucket).
3. The milking parlour for use with yarded cows; and movable milking bail for herds at pasture.
4. Round-the-shed (RTS) milking involving direct passage of the milk from the teat-cups into a milk pipeline along which it is delivered direct to churn or bulk-tank. The milkline should have few right-angle bends to avoid 'churning action' which can result in taint development.
5. Rotary milking unit; in which the operator is normally stationed in one position throughout milking.

Milking Parlours

The principle of the parlour is that the cow walks from the soiled area, contaminated by manure and dust, into a highly hygienic area to deliver her milk and, often, to receive her concentrated feed. The operator has to do very little walking about and carrying but instead concentrates his efforts on washing the udder and teats and milking the cows as they present themselves. The milk is delivered under vacuum into bulk tank or churn direct from the teat-cups.

There are five main types of parlour installations:

1. Abreast single-level parlour in which, as in the original milking bail, cows stand side-by-side for milking. They walk through the stall back to their accommodation.
2. Abreast two-level parlour which is similar except that the cows walk up on to a raised stall 350-430 mm above the floor level, on which the operators stands, who therefore has to stoop very little to put on and take off the units.
3. Tandem-type parlours where the cows stand side-on in

line to the operator's pit, exposing conveniently the udder and teats for the operator to wash and to attach the milking units. As each cow is dealt with and released another takes her place in the stall so that milking is a continuous process.

4. The "chute" parlour is especially suitable for narrow buildings. The cows, however, are let in and out in batches. This layout is similar to the tandem parlour, is considerably cheaper to install but suffers the disadvantage that the batch cannot be released and replaced until the slowest milking cow has been dealt with in each batch.

5. The herringbone parlour where the cows stand in groups of 4, 8 or more at an angle to the operator's pit without intervening stall divisions, exposing only flank and udder to the operator.

6. Rotary parlours. These parlours rotate around the operator and incorporate either the tandem or the herringbone principle. The complete rotation is completed in 6-7½ minutes, movement can be halted by the operator while cows move into their stalls and move out. Units of 8, 12 and 18 stalls are avail-

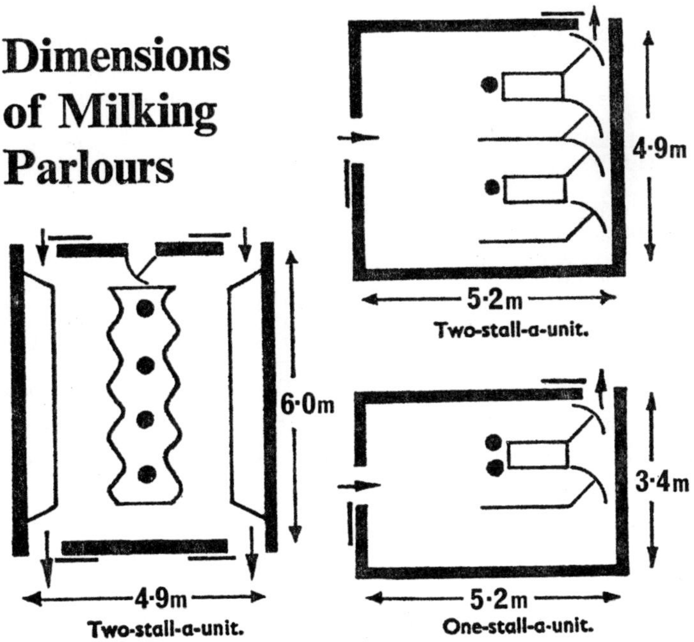

Herringbone and abreast parlours.

able and throughput ranges from 85, 90-100 and 150 cows per hour, and these are operated by one, one and two men respectively. The trend of development is along the lines of automating cluster-removal, cow exit and entry control, udder-washing and feed control. Also safety-stop devices are being incorporated to avoid accidents.

Since the first rotary parlour was installed in Britain in 1969 they have increased in popularity because, while they increase output per man, they also provide a greater degree of comfort to the man and the cow. It is probable that some of the refinements built into design will improve milking technique and precision cow management—e.g. low level milking units, pipe line and calibrated milk jars.

Size of Herd

Total milking time should not exceed 2 hours in the morning and within this period 10-15 cows can be milked by one unit. The number of units a man can handle without being overtaxed or under-employed in milking is governed largely by the 'work routine' which in most tandem and abreast parlours compared with the N.I.R.D. Herringbone can be summarised:

	Tandem and Abreast (*seconds*)	*NIRD Herringbone* (*seconds*)
Let in and feed cow	30	6·6
Take fore-milk and wash udder	25	7·2 and 12·0
Attach teat-cups	20	16·2
Cow being milked by machine		
Machine strip and dip teats	15	4·2 and 4·8
Record	10	6·6
Let cow out	10	4·2
Emergencies and idle time	10	4·8
	120 seconds	66·6 seconds

The work-content per cow in tandem and abreast parlours is therefore two minutes and this puts the maximum efficient throughput at 30 cows per hour per man. Of course, if the routine is modified, the throughput changes.

In the operation of the herringbone batches of 8 or more cows are admitted and, later, released at once. Since the cows are closely packed together, the time spent on washing the udder and machine-stripping involves minimum movement by the operator and the 'work routine' takes less time. It can be reduced to $1\frac{1}{2}$ minutes and consequently the maximum throughput per hour is 40 cows. In rotary parlours many tasks are automated and the cows rotate so that the movement by the operator is at the absolute minimum.

LAYOUT OF PARLOUR

		Up to 30 *cows*	30-40 *cows*	40-60 *cows*	60-250 *cows*
Size of Herd					
Type of Parlour		Tandem or Abreast	Tandem or Abreast	Herringbone	Rotary, Tandem or Herringbone
No. of units per operator	winter milk	2	3	4	5-9
	summer milk	3	4	5	5-9
No. of stalls per operator	winter milk	3	4	10	5-9
	summer milk	4	6	8	5-9

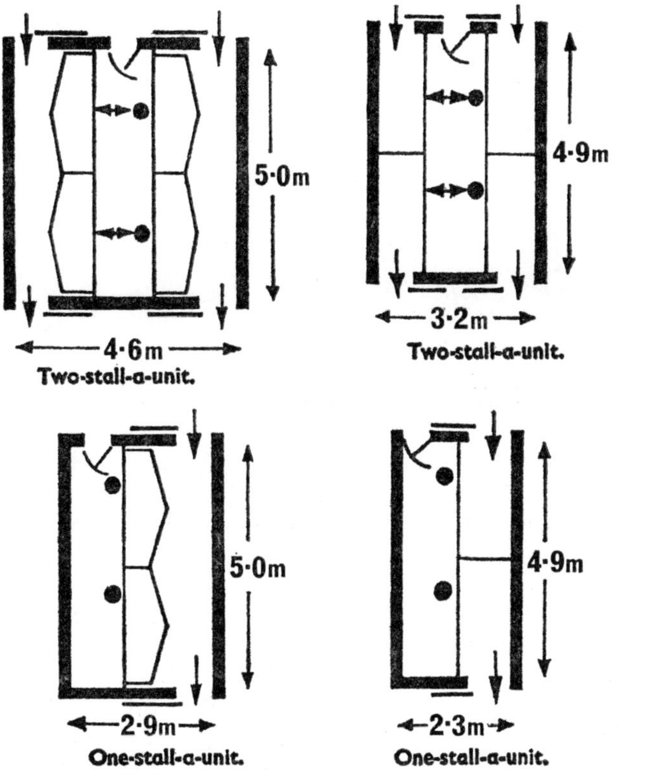

Double and single tandem and chute parlours.

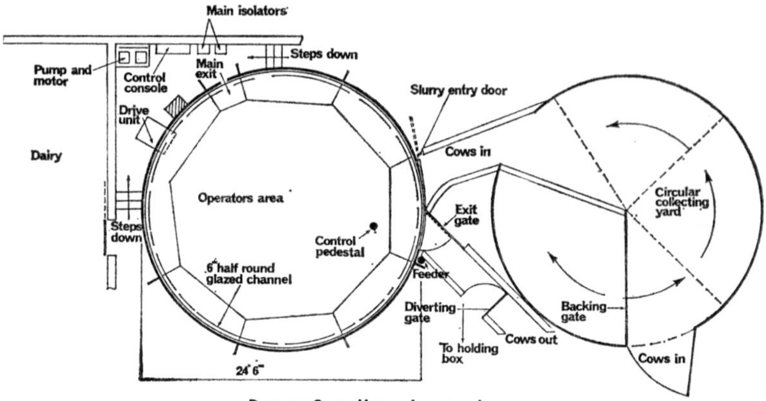

Rotary 8-stall tandem parlour.

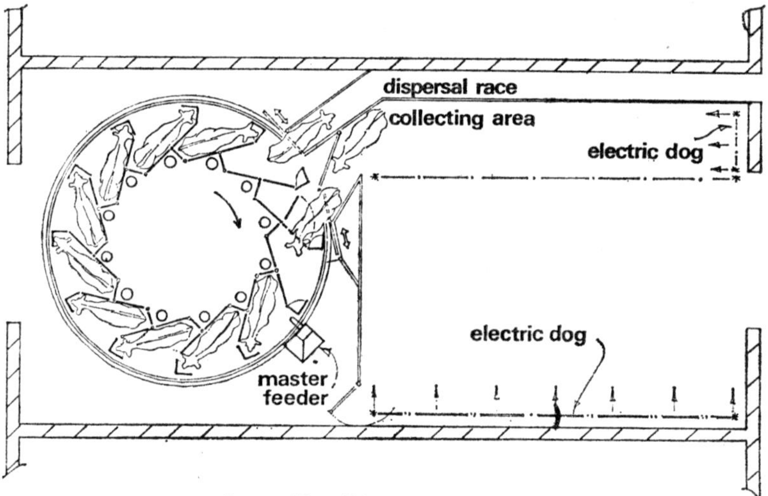

Rotary 12-stall herringbone parlour.

Marshalling arrangements are shown with both parlours.

BULK TANKS

By 1968 some 5,900,000 litres of milk a day were being handled in bulk in England and Wales.

These tanks are filled during milking direct from the milk pipeline. Cooling takes place in the bulk tank which consists of an inner stainless steel tank surrounded by water or refrigerated coils with an outer insulated shell. The milk is cooled to 4°C and stored until collected in bulk by road tankers. These vehicles are very heavy and adequate roads are essential.

In practice, the bulk tank is well worthwhile installing, particularly in view of the price incentives offered by the Milk Marketing Board, and also because the chore of churn sterilising and supervision during milking makes constant demand on the attention of the operator and afterwards there is the work involved in churn-labelling and handling.

Slurry Disposal and Pollution

On light land it is possible to dispose of the thin slurry from cows housed on slats, in cubicles or cow kennels in mobile tankers from which the excreta is spread. Some spreaders will cope with muck or little straw. In wet seasons, however, problems do arise.

An alternative is organic irrigation. The liquid excreta, straw excluded, is stored in collecting tanks and agitated adequately by paddles before entering the irrigation pipeline through which it is conveyed to rain guns for disposal over the land.

Storage capacity required is 640 litres, 0·7 m^2 per cow if the tanks are emptied weekly. Each cow will provide manurial elements equivalent to 150 kg of 20:10:20 fertiliser. A herd of 40 cows, would provide 25 mm of slurry on one hectare every week and the long-term effect of this application to grassland is not known under our conditions.

Alternatively, the slurry is stored in lagoons or 'turkey nests' until it has dried out sufficiently and, later in the summer, when the land is dry, it is spread. Turkey nest storage, as the name implies, involves the temporary erection of a wall of solid manure around the area in which the slurry is stored, perhaps to a depth of 1·2 m in a dry part of a nearby field.

Dairy farmers are becoming annually more and more aware of the public alarm and local bylaws regarding pollution of the environment. River Boards are actively examing water courses for pollution by excreta, silage and other farm sources, particularly those arising from fertilisers.

CHAPTER V

ROUTINE TASKS

IDEALLY, feeding programmes are based on calculations of ME and DCP required by various classes of cattle as revealed in Chapter I. However, traditional practices of home-mixing concentrated foods have achieved good results and they are included in this edition because of their simplicity and usefulness.

Where home-mixing of concentrated foods is carried out, the proportions in which home-grown foods such as oats and barley are mixed with purchased oil cakes such as linseed, cotton seed, or groundnut cakes must be related to the class of stock being fed and to current market prices.

Class A: Calves and Dairy Cows

Young calves, from weaning up to eight months old, and dairy cows in milk or being steamed up, require a concentrate mixture in which there is adequate DCP (dry protein)—unless bulk foods of a high quality such as first class hay or silage are being fed.

Such a mixture can be made up from the following guide:

Group I

(Foods already balanced for milk production)

Compound cakes
(albuminoids 18-20%)
Palm kernel cake or meal
Coconut cake
Weatings
Bran

Maize gluten feed
Brewers' dried grains
Dried grass (16-18% protein)
Dredge corn with at least 30% peas and/or beans

Group II

(Mix 1 part with 1 part of any food in Group VI)

Linseed cake
Peas and beans
Distillers' dried grains
Malt culms
Undecorticated cotton cake
Sunflower seed cake
Grain balancer (cake or meal)

Group III

(Mix 1 part with 3 parts of any food in Group VI)

Undecorticated groundnut cake
High-protein cake or meal
Maize gluten meal

Group IV

(Mix 1 part with 4 parts of any food in Group VI)

Decorticated groundnut cake or meal
Decorticated cotton seed cake or meal
Soya bean cakc or meal

Group V

(Mix 1 part with 6 parts of any food in Group VI)

White fishmeal
Feeding quality meat meal
Dried yeast (in limited quantities only)

Group VI

(Cereal foods)

Maize meal
Maize germ meal
Crushed wheat, rye, barley or oats
Dredge corn with less than 30% peas and/or beans
Flaked maize
Locust beans
Tapioca meal
Dried sugarbeet pulp
Molasses
Dried potato products

Thus a suitable mixture might be as follows, assuming protein cake in the form of decorticated groundnut cake has been purchased:

Crushed oats 2 parts Crushed barley 2 parts	} Group VI, 4 parts
Decorticated groundnut cake 1 part	} Group IV, 1 part

CLASS B: GROWING YOUNG STOCK

For young stock of from 8 months to 24 months, including beef and dairy cattle, add 20 per cent (one part in five) of cereal food to a balanced ration for Class A stock. Thus:

Crushed oats 3 parts Crushed barley 2 parts	} Group VI, 5 parts
Decorticated groundnut cake 1 part	} Group IV, 1 part

CLASS C: FATTENING CATTLE

Fattening cattle should have 40 per cent (two parts in five) of cereal food added to a balanced ration for Class A stock. An example is:

Crushed oats 3 parts Crushed barley 3 parts	} Group VI, 6 parts
Decorticated groundnut cake 1 part	} Group IV, 1 part

Some further examples of this method are as follows:

Foods	*Dairy cows and calves*	*Young stock (20% more cereal)*	*Fattening cattle (40% more cereal)*
A. Oats	4	6	8
Beans	4	4	4
Palm kernel cake	2	2	2
	10 kg	12 kg	14 kg

B. Oats	3	4	5
Barley	2	3	4
Grain balancer cake	5	5	5
	10 kg	12 kg	14 kg

C. Oats	4	4	*5
Barley	3	4	5
Palm kernel cake	1½	1½	1½
White fishmeal	½	½	½
Decorticated groundnut cake	1	1	1
	10 kg	11 kg	13 kg

*This ration is especially suitable for baby beeves and young bulls.

The inclusion of a complete mineral mixture is recommended in all home-mixed concentrates, at rates dependent on manufacturer's instructions. To exceed normal additions is more likely to do harm than good in that excess of minerals in the diet can upset the mineral balance in the diet as a whole and render some trace elements (e.g. cobalt) still less available.

Where white fishmeal is employed the recommendation is to add minerals at half the rate.

A home-mixed mineral mixture suitable for inclusion at the rate of 1·12 kg per 50 kg of mixed concentrates is:

Powdered rock salt or crude salt	1 kg
Ground limestone	½ kg
Steamed bone flour	1 kg

Where kale is being extensively fed to dairy cows an iodised mineral mixture is advised, and where lucerne or dried grass are being fed in quantity a mineral mixture of high phosphorus content is recommended. Mineral mixtures fortified with magnesium are also available now wherever cases of lactation tetany or hypomagnesaemia have occurred in the herd.

Method of Mixing

In the absence of food-mixing machinery, spread the coarsest textured food on a hard floor first, followed by each food in consecutive layers until the total mix is heaped on the floor, then turn twice with a shovel. Minerals should be thoroughly mixed with one of the ingredients first and then incorporated in the main mixture.

Cereals such as oats and barley are better fed rolled than crushed or hammer-milled. In the former state palatability is greater, and less waste occurs from blowing.

Beans and peas are best fed coarse ground or kibbled.

Oil cakes can be purchased either in meal form, or as cubes, or in kibbled form for mixing with straight feedingstuffs. Cattle—in contrast to pigs—find relatively coarse-textured concentrate mixtures more palatable than meal.

Certain foods are useful to give a favourable texture to meal mixtures; such foods are bran, brewers' dried or distillers' grains and malt culms. Other feedingstuffs are noted for their laxative or costive properties; when they are used it is important to make balancing modifications in the concentrate mixture to promote normal functioning of the bowels.

A mash made from soaked beet pulp or scalded bran (using *hot* water) is often useful as a tonic feed. The addition of molasses will also add to the palatability of a mash feed. Do not use too much water, the mash should be just crumbly in texture. 0·3 litre of molasses with 2 kg bran or beet pulp is a common recipe.

How to Feed

The method of rationing concentrates depends very much on building layout.

One way is to use a food barrow from which the requisite allowance of concentrates for each beast is measured out at feeding time. This requires the allowance per beast to be available either from feeding scales on the barrow or chalked up above each trough.

This method, with dairy cows, may lead to careless feeding if it is done while milking is in progress; it tends to overload the routine at milking time.

Another method is to weigh out the concentrates into individual buckets for each beast. This can be done between feeding times; it gives greater accuracy and spreads the work and is recommended for high-yielding cows or cattle being fattened for show or sale purposes.

There are also the automatic or shovel-feed systems for use in milking parlours.

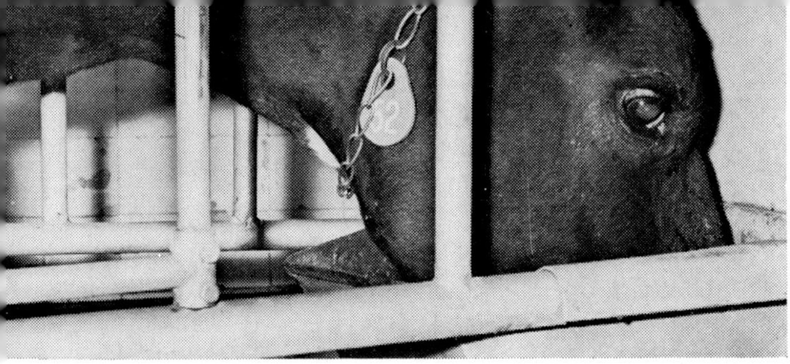

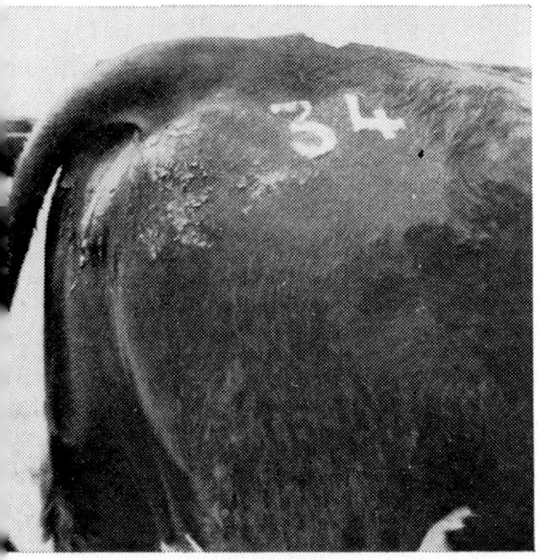

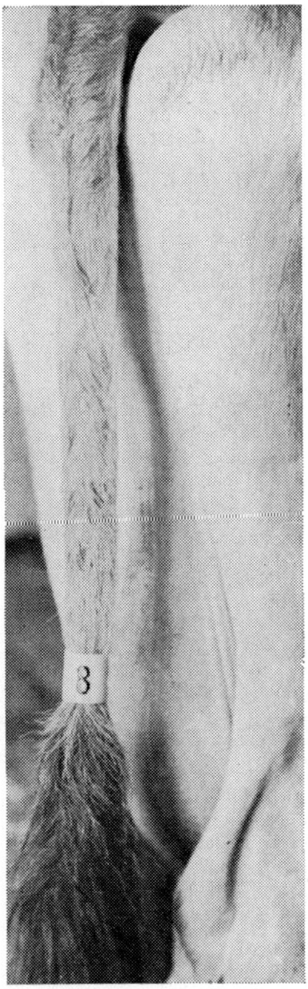

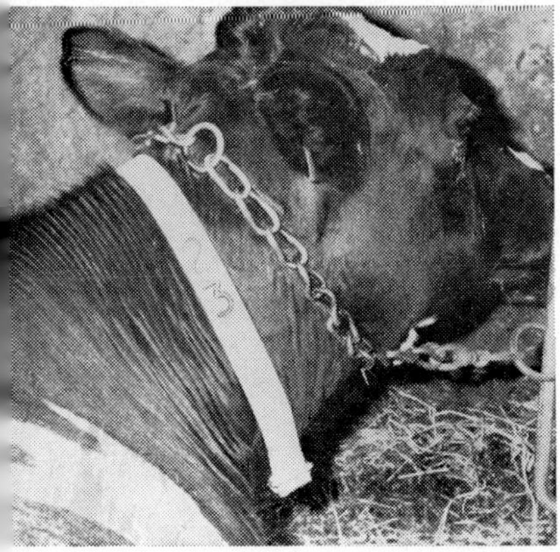

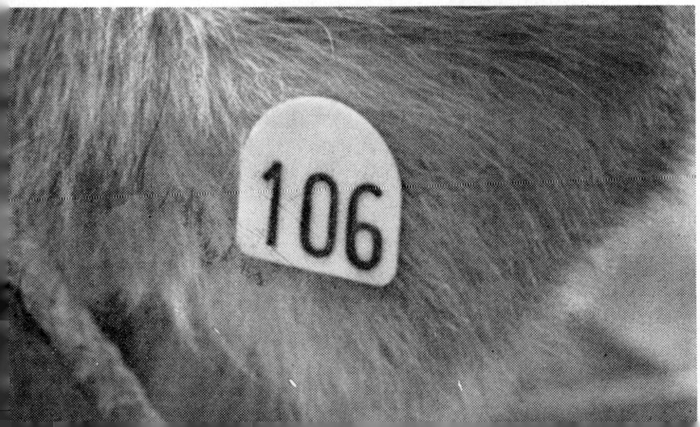

Cow identification is one of the problems with cows kept in large numbers. The possibilities include ear, neck, tail and ankle tags. Promising development is 'freeze branding' (second picture down on left).

Milk recording is one of the essential management aids but can be time-consuming and cause disruption of milking routines. Ways of getting over these problems include tape recording of yields (top), recording sheets on a monorail arrangement (centre), and flow recording in pipeline plants (bottom).

Essential parlour fixture is some form of feeding and breeding record chart. This see-at-a-glance board gives service and return dates and amounts of concentrates to be fed, brought up to date once a week.

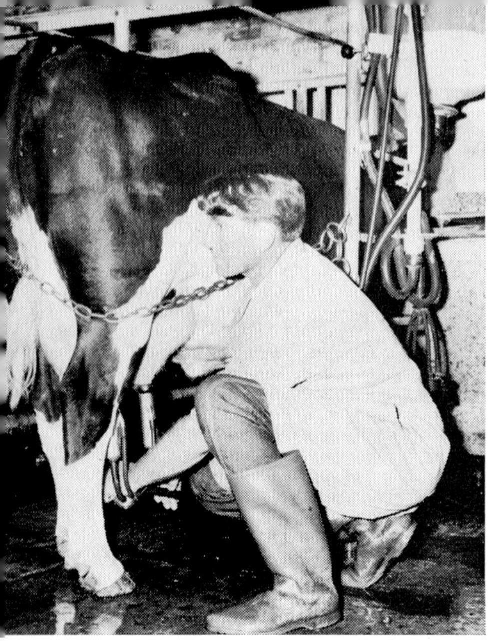

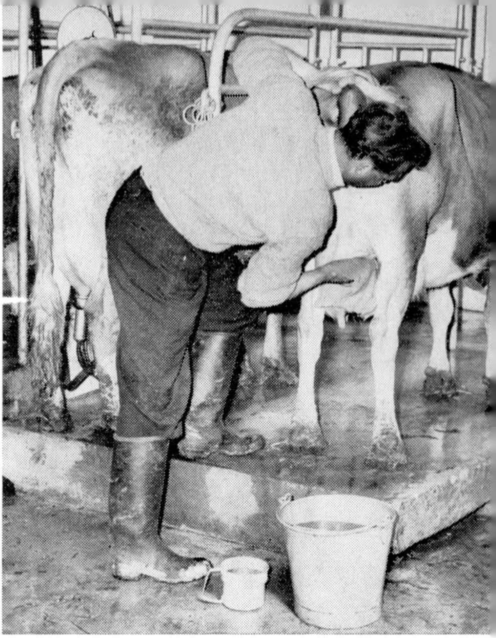

These four pictures showing working positions in the main types of milking parlour—the stooping involved with a single-level abreast (top left), alleviated by raising the cows on a platform (top right), and eliminated by putting the herdsman into a pit in the tandem and herringbone parlours (bottom left and right).

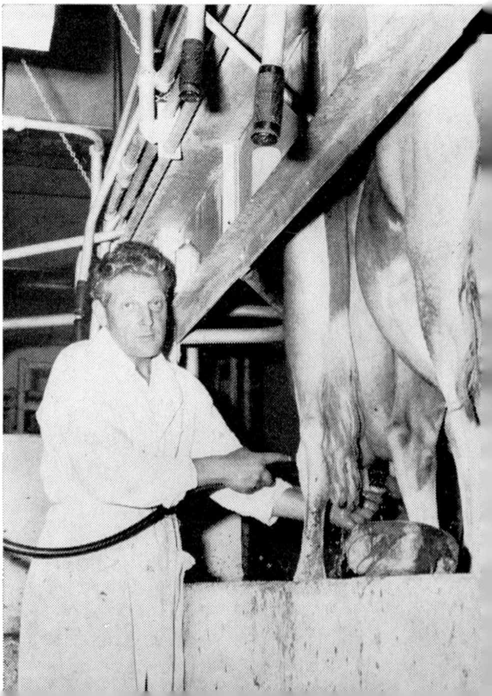

The Ruakura milk flow indicator helps to avoid overmilking by providing a visual indication of milking coming to an end—top, machine stripping stage; centre, prepare to remove cups stage; bottom, take cups off stage.

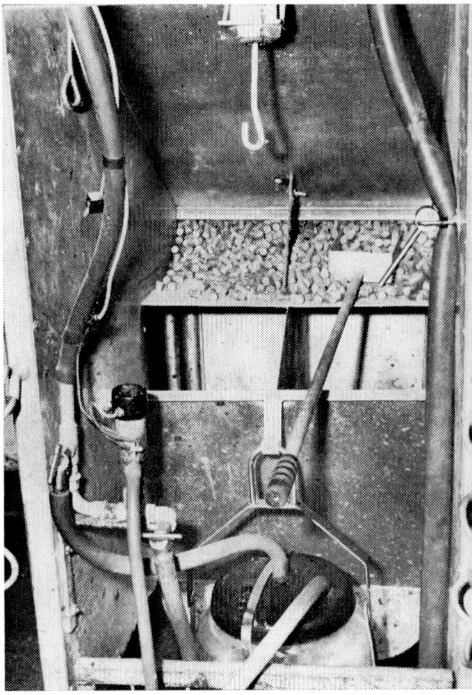

Feeding according to yield can be based on direct weighing with a spring balance (top), by volume using a scoop holding a known approximate weight, (centre) *or from a hopper releasing a set amount at each turn of a handle (bottom).*

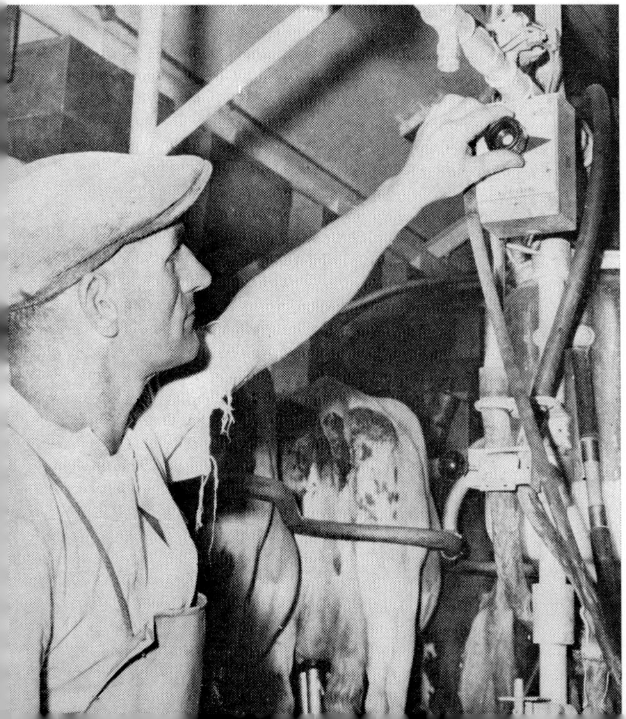

Latest plants incorporate virtually fully automatic feeding. This herringbone has vacuum-operated dispensers filled by auger conveyor to eliminate all feed handling (above). The herdsman determines the amount to be fed to each cow by setting a pointer on a dial (left).

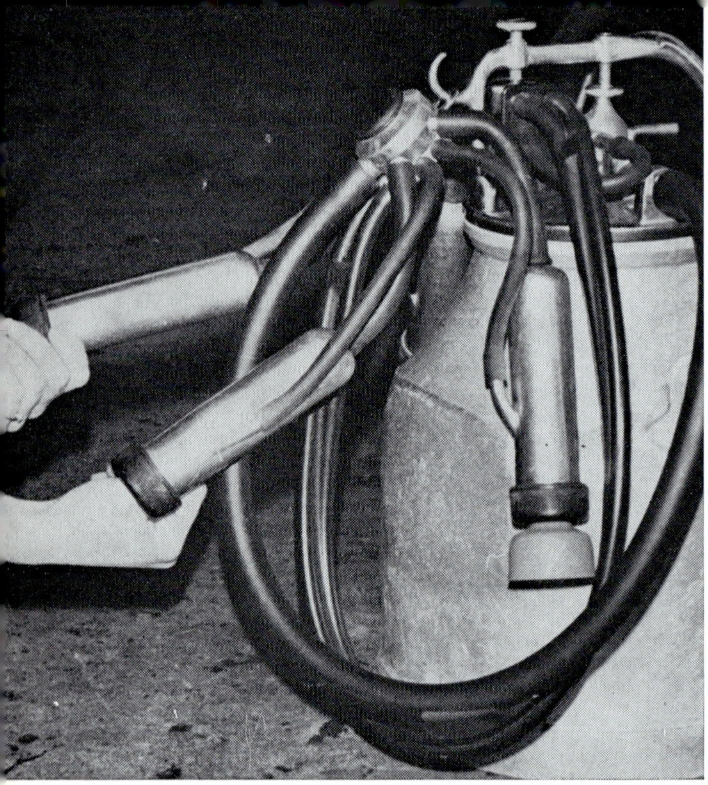

Many milking plants are allowed to operate at below optimum efficiency. Here pulsation rate is being checked by blocking three teat-cups and inserting a thumb in the fourth.

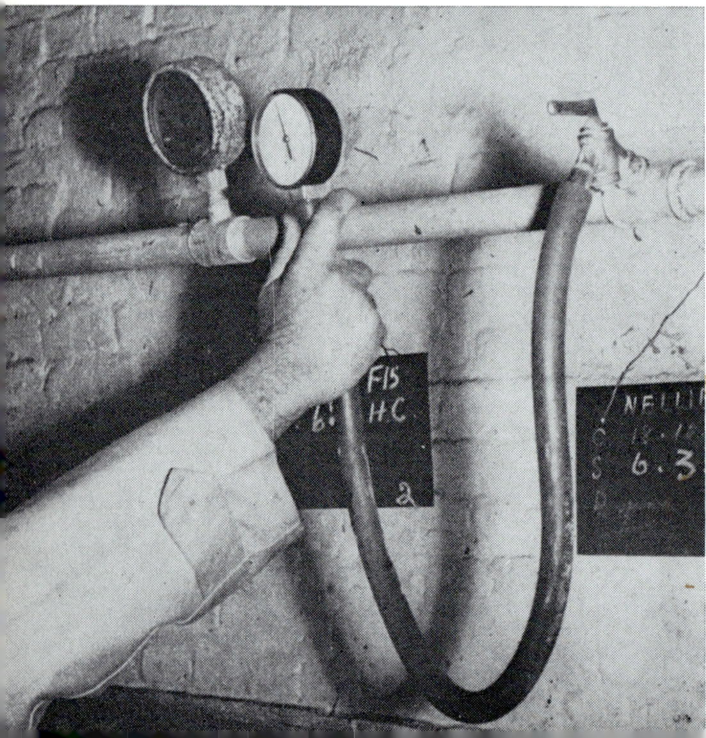

Vacuum control is the 'brain' of a milking plant, so regular checks on vacuum efficiency are essential.

Finally, where concentrates are auger-fed to groups of cows, it is necessary to group cows of similar yields together.

Identification of Cows

It is obviously essential in any rationing scheme to be able to identify individual cows or groups of cows at similar yield levels.

In cowsheds this is no problem, as cows are tied up in their individual stalls which can be numbered. With yarded cattle, four methods are available:

1. Tattooing the number of each cow on her rump with a cold caustic paste—the branded area must be kept clipped at two-monthly intervals to retain legibility. With single level parlours this method is quite suitable, but with two-level operation the brand marks may not be visible from the operator's pit.

2. Use of neck straps with a numbered disc attached, or a large disc attached to the metal ear clips now commonly in use for identification under the Attested Herds Scheme. In herring-bone parlours this method is not so effective as only the hind-quarters of the cows can be seen.

3. Use of plastic coloured tapes which can be tied on the cows' tails, identifying different yield groups by the colour of the tape—e.g. red for cows giving 20 litres, blue for cows giving 15 litres and yellow for cows giving 10 litres.

4. Cryogenic cold branding, or "freeze branding". This technique is based on the use of an extremely cold branding tool applied to the part of the cow appropriate to the necessity for its visibility—for example, in two-level parlours on the lower thigh.

Clean Milk Production

The immediate responsibility for preventing bacterial contamination of milk produced for liquid consumption rests with the herdsman. The responsibility for providing adequate facilities rests with the herd owner.

Conditions governing milk production are laid down in the Milk and Dairies Regulations, 1959. The essentials can be summarised as follows:

1. The cows must only be milked in premises (including milking bails) approved by the Ministry of Agriculture, such premises must be kept clean and adequately ventilated.

2. Before milking, each cow should be cleaned of any gross contamination with dung, using a curry comb and dandy brush. Then, immediately before milking begins, the udder and teats should be washed in luke-warm water 49° C and wiped dry; milking should follow within two minutes of washing. The milker should wear clean white milking overalls,

3. Immediately after milking the milk should be conveyed to the milk room or dairy, filtered and then adequately cooled—to below 10° C if possible. If conveyed in pails, the pails should have covering lids.

4. Once milking is finished all equipment coming into contact with the milk should be thoroughly rinsed in cold or luke-warm water, before any milk can dry on the surfaces of buckets, pipelines and so on.

5. Final cleaning and sterilisation is then performed by one of four methods:

(a) Washing with a hot detergent solution made up according to maker's instructions—usually 5 g in 1 litre—to remove fat, followed by sterilisation with steam in a steam chest at a minimum temperature of 96° C for 10 mins. Milk tubes and teat-cup clusters are sterilised separately on steam jets for 3-5 minutes.

Equipment needed—steam raising equipment; two-compartment wash-up trough; steaming chest with steam jets fitted to racks (for storage);—table for dismantling and assembling teat-cups.

(b) Washing with hot detergent solution, as above, followed by immersion of all equipment in a hot hypochlorite solution made up according to maker's instructions—usually 1·25 g in 1 litre—for at least two minutes. Then allowed to dry off on suitable storage racks. The teat-cup cluster can be stored immersed in hypochlorite solution (wet storage method).

Equipment needed—hot water boiler; two-compartment wash-up trough; racks for storage of teat-cup clusters; table for dismantling and assembling teat-cups.

(c) Immersion cleaning—applicable to direct-to-churn milking systems. After the preliminary cleaning, teat-cups are released from the liners, rubber stopper removed from the teat-cup cluster claw, and teat-cup clusters, milk tubes and churn lids packed in a wire basket.

The basket is then immersed completely in a 55 litre plastic drum containing 45 litres of 2 per cent caustic soda—a water softening agent, EDTA, is added to the caustic soda at 60-170 g according to hardness of the water. The milking machine components are left in soak between milkings.

To prepare for milking, the basket is removed, the caustic soda solution swilled off with a hose and the components unpacked into a warm solution of hypochlorite—30 g in 50 litres—before re-assembly.

The caustic soda solution is renewed each month.

Equipment needed—hot water boiler; plastic container and immersion basket; two-compartment wash-up trough with draining board; teat-cups and churn lids of stainless steel.

(d) Circulation cleaning is the most recent development with pipeline or auto-recorder plant. By using either the milking machine pump or an auxiliary milk pump the whole plant is washed by circulating the various washes within a closed circuit.

Trials carried out by the NAAS in which the author took part emphasised the importance of a correct routine, rigidly observed, in maintaining high hygienic standards in the milk produced. A satisfactory routine can be summarised as follows:

Daily Routine

(1) After milking brush off all adhering dirt from the teat-cup clusters and milk tubes.

(2) Flush whole system with cold water to remove milk residues.

(3) Remove teat-cup clusters to dairy, attach to suck-up points and immerse in tank containing hot (71° C as minimum) detergent/hypochlorite solution—250 g detergent and 125 g hypochlorite in 50 litres water. Connect water line to milk pipeline to complete the circuit. Allow the solution to circulate for a minimum of 10 minutes—temperature of solution should not fall below 38° C at completion.

(4) Circulate cold water to remove all traces of detergent/hypochlorite.

(5) Immediately prior to milking, circulate a hypochlorite rinse—31 g to 50 litres of hot water 54-66° C, or 62 g to 50 litres if cold water used. This pre-milking rinse is most important in hot weather.

Weekly Routine

Once a week the plant should be flushed with milk-stone remover—280 g to 50 litres cold water. The milk-stone remover solution can be used for six weeks before renewal. Teat-cup liners should be removed and left in soak in 5 per cent caustic soda solution on alternate weeks to remove fat.

Equipment needed—hot water boiler; suck-up points in dairy and 50 litre container; table or draining board for dismantling and assembling teat-cups.

Both immersion and circulation cleaning are labour saving and cheaper than the traditional use of steam. But circulation cleaning is unreliable unless special care is taken to prevent build up of *E. coli* contamination on rubber gaskets and pipe-line connections, which should be reduced to the minimum in plant design.

To avoid using both detergent and sterilising solutions, the NIRD has developed a once-through system of cleaning based on the use of nitric or sulphamic acid.

The teat-cups are attached to jetters closing the circulation line, and acidified boiling water is then circulated for 2-3 minutes—the temperature must not fall below 77° C to prevent any deposition of milk-stone—followed by boiling water only for 2-3 minutes. The total volume of water needed per milking point is approximately 13 litres.

Cleaning of bulk tanks can be done by the use of approved detergent and sterilising agents, as described earlier in circulation cleaning, but the NIRD has also demonstrated a once-through automatic cold cleaning system, using an approved iodophor solution as a combined cleaning and sterilising agent.

The technique involves:

(a) Rinse with cold water.

(b) Spray for $\frac{1}{2}$ minute with an iodophor wash (190 g in 50 litres).

(c) Leave for 20 minutes.
(d) Rinse with cold water (115 litres) sprayed for $1\frac{1}{2}$ minutes.

Equipment Maintenance

Lastly, one other routine job—the maintenance of dairy and milking machine equipment. Here is a guide:

Daily

Check level in water gauge and fill boiler if necessary before steam is withdrawn from the boiler—this must be done daily.

Weekly

Check the oil level on the milking machine pump and oil the electric motor, or check oil level in petrol engine driving the pump at least weekly.

Check vacuum in pipeline, ensure all taps are air-tight and adjust release valve if necessary.

Check pulsation rate with a stop-watch. Adjust individual pulsators unless automatic control being used.

Change teat-cup liners. Use two complete sets, storing one set in 5 per cent caustic soda whilst the second set is in use. Make sure the teat-cup liners are under tension when fitted—stretched liners cut to the required length, moulded liners stretched by applying tension and held in position by corrugations on the small milk tubes.

Every three months

Where hard water is being used, unless a water softener is installed, the boiler will need to be de-scaled at three-month intervals with an acid de-scaling solution obtainable from dairy engineers. Dairy equipment itself should be de-scaled by the use of de-scaling compounds based on phosphoric acid, proprietary brands being available from dairy supply companies.

Antibiotic Residues

At least 48 hours must have lapsed after a cow is treated with antibiotics before the milk is eligible to be sold.

All milk that passes through the MMB is tested for residues. The first test failure is notified to the producer so that he can take appropriate action. The second failure results in a warning letter and a third failure brings a price reduction on the milk tested.

CHAPTER VI

CALF REARING

THE health and future well-being of the calf is laid by good feeding and management from birth to six months of age.

Methods of rearing vary, but common to all methods is the need to ensure that the young calf receives an adequate supply of colostrum which has the property of building up resistance to infectious conditions of the bowels often referred to as "scours".

Value of Colostrum

Colostrum is more easily digested than ordinary milk, is rich in vitamin A and has laxative properties which help the bowels of newly-born calves to function normally and to expel the foetal dung.

If cows are milked before calving (pre-partum milking as it is called) their colostrum should be saved if possible by refrigeration for the calves at birth.

Where Johne's disease is known to occur in a herd, the colostrum should be pasteurised (heated to 60-63° C for half an hour) before being fed.

It is now believed that much of the infection with Johne's disease is contracted during the suckling period.

Where colostrum is not available a substitute can be made up as follows: whip up a fresh egg in 0·9 litres of milk (preferably from a freshly-calved cow), add 0·25 litres of boiled water, 1 teaspoonful of cod-liver oil and 1 dessertspoonful of castor oil. Mix well with an egg whisk and give as one feed at 38° C —temperature is important.

Veterinary opinion suggests that where calf scours are at all common the calf should also be given an injection of blood serum as soon after birth as possible—thus transferring the protective antibodies from the blood of the cow to the calf, as normally happens where true colostrum is fed.

Importance of Warmth

If adequate colostrum is important, so is adequate warmth. Young calves are easily chilled, and up to three to four weeks of age lack the heat-producing mechanism of the cud-chewing adult. They should, therefore, if housed, be kept well-bedded in warm, dry quarters for the first month of age.

Methods of Calf Rearing

1. *Single Suckling*

The simplest and easiest method of calf-rearing is the single-suckling system, each cow rearing her own calf.

This is most common in beef production; it is the method most expensive in overheads, but is justified where first-class suckled calves are the aim and where the cows can be out-wintered and cheaply fed.

Calving in such suckler herds is usually timed for February/March so that the cows calve six to eight weeks in advance of the grazing season.

When the grass does come the calves are old enough to take the flush of milk when their needs for growth are at a maximum; later calvings may mean too much milk for the calf and severe scouring—unless the cow is kept on a bare pasture for a time.

Beef herds in hill country which are housed in cubicles in winter are ideally arranged to calve in September/October as explained on page 159.

2. *Multiple Suckling*

A development from the single-suckling system is found where, with better food supplies for the cows and some buildings available, each cow can be expected to rear seven or eight or more calves by a continuous process of multiple suckling.

At her peak of lactation a good nurse cow may suckle four calves for, say, three months, then three more calves, and finally two—making nine in all. But few nurse cows achieve this unless they are specially selected for heavy milking ability. A nurse cow is usually a cull from the dairy herd and may have lost a quarter, or is a slow milker, but if she is a good mother and makes a "fuss" of her foster calves, she fits this system of calf-rearing very well.

Controlling Nurse Cows

Some control of nurse cows is needed with multiple suckling. They can be tied up in a byre or cowshed, and the calves brought in for suckling twice a day. Or the cows can be tied in loose boxes in which the calves are normally housed; the rest of the time the nurse cows can spend out-of-doors or in a yard.

Improvised accommodation can be used for this system without great expense, provided the calves are given reasonable shelter and a dry bed.

Multiple suckling is a system that fits in very well on arable farms where rearing begins in the autumn. One batch of calves is reared before Christmas, a second batch by March, a third batch during the early part of the grazing season—during which the nurse cow can be expected to fatten if she is not kept for breeding purposes.

The biggest drawback to this system—where enough calves are not bred on the farm—is to find suitable calves when wanted. This is most difficult in the spring; hence autumn suckling should be considered where calves have to be purchased. In any case the cost of and the mortality rate in such calves is usually less than in spring-born calves.

To introduce a calf to its foster mother requires patience, and the following technique is recommended:

First ensure the cow is tied up and unable to hurt the calf by kicking. If she is a shy mother, spanning her hind legs above the hocks with a rope in a figure-of-eight fashion will effectively restrain her.

Then insert a finger in the calf's mouth and guide the calf into the normal suckling position. Entice its head towards the teat not by pushing on the top of the calf's head but by leading it with the finger being suckled until the cow's teat can be substituted for that finger, when suckling will usually commence in earnest.

Do not push on the calf's hindquarters and hope it will find a teat to suckle. It may do so, but many nurse cows with pendulous udders require a more subtle technique.

Calves that have been bought in should not be allowed to gorge themselves when freshly introduced to their foster mother, even if it means milking out the cow by hand for the first couple of meals.

And when changing over from one batch of calves to the next, allow the new arrivals to suckle first. Then let the older calves finish off the cow and wean themselves in the process.

GUIDE TO SINGLE AND MULTIPLE SUCKLING SYSTEMS

	Method of Rearing	
Feeding at	*Single Suckling*	*Multiple Suckling*
One week	Suckling dam	Suckling nurse cow
One month	Suckling dam	Suckling Hay and concentrates offered as eaten
Three months	Suckling dam Concentrates offered in creep and/or grazing	Suckling 1·3 kg hay; 1·3 kg concentrates
Four months	Suckling dam Concentrates offered in creep and/or grazing	Suckling ceases 1·3 kg hay; 1·8 kg concentrates
Six months	Suckling ceases 1·8-2·7 kg concentrates. Hay as eaten or grazing	2·7 kg hay; 2·3 kg concentrates or grazing
Liveweight gain per day from birth	0·8-1·0 kg	0·6-0·7 kg
Approx milk consumption	1130-1360 litres	410-460 litres

3. *Pail Feeding*

The third, and probably in terms of numbers reared, the most important method of calf rearing is by pail feeding on limited quantities of whole milk supplemented by proprietary foods known as milk equivalents or milk substitutes.

This method requires more specialised accommodation in the way of calf pens (preferably for individual calves) and is more exacting in labour demands, but it does enable a considerable saving in food costs to be made compared with the whole milk feeding as in the previous methods.

The level of nutrition is lower with this method than the others and is more suited to rearing dairy-type calves but can be raised to give results comparable with multiple suckling.

In pail feeding it is most important to see that meal-times are regular and as evenly-spaced as possible, that all feeding buckets are kept scrupulously clean and sterilised at least once a day, and that all-milk equivalent or milk-substitute foods are correctly mixed and fed at blood heat 38° C.

The mixing of calf foods is best done in the calf house or adjoining food store with a convenient supply of hot water (as from a small electric hot water boiler) and with a thermometer available to check temperatures.

It is necessary to teach calves to drink from a bucket at first. This again requires patience. Stand to the side of the calf's head. Hold its milk feed in front of it in a shallow bowl or a small bucket gripped in your left hand. Put your right arm over the calf's head and insert one finger of the right hand in the calf's mouth.

As soon as the calf sucks, lower its head into the milk and gradually remove the finger.

Holding the bowl at knee height will allow the calf to drink more readily as it tends to suckle at that height and will resent any attempt to push its head into a bowl or bucket at ground level.

Whilst being fed on whole milk the dilution of milk with 25 per cent of added water will reduce the risk of digestive scour, as will the avoidance of giving too much milk at any one feed.

Rearing on the Bucket

Dairy Calves

Either of two methods of calf rearing by bucket feeding can be employed:

Firstly, the older or traditional method which involves feeding the calf on whole milk until 3-4 weeks of age and then gradual replacement of milk by milk substitute feeds, weaning at 12 weeks or so.

Secondly, the more recently introduced early weaning system where milk is cut out after the colostral period of 4 days, followed by up to 5 weeks on milk equivalent feeds, then weaning on to a diet of dry feeds plus water.

Suitable guides to the feeding of calves on the two systems are:

The Traditional System

The First Three Weeks

Breed	*Whole Milk* litres	*Water* litres
Jersey	2·3→3·4	0·6
Ayrshire and Guernsey	3·4→4·5	0·8
Dairy Shorthorn and Red Poll	4·5→6·2	1·1
Friesian	5·7→6·8	1·4
South Devon	6·8→8·0	1·7

Remarks

Colostrum essential, milk given in three feeds if possible for first week.
Level of feeding increases as shown over the period.
After three weeks of age hay is offered to the calves with dry concentrates as eaten.

From Fourth to Eighth Week

Breed	*Whole Milk* litres	*Milk Substitute* litres	*Concs* kg	*Hay*
Jersey	3·4→0	1·1→5·7	0·1→0·5	ad lib
Ayrshire and Guernsey	4·5→0	1·1→5·7	0·1→0·7	ad lib
Dairy Shorthorn and Red poll	5·7→0	1·1→5·7	0·1→0·9	ad lib
Friesian	6·8→0	1·1→5·7	0·1→1·1	ad lib
South Devon	8·0→0	1·1→5·7	0·1→1·4	ad lib

Remarks

Change-over from whole milk to milk substitute made gradually as shown.

From Ninth Week to Weaning at Twelfth Week

Breed	*Milk Substitute* litres	*Concs* kg	*Hay* kg
Jersey	5·7→0	0·5→1·1	Up to 1·4
Ayrshire and Guernsey	5·7→0	0·7→1·4	Up to 1·6
Dairy Shorthorn and Red Poll	5·7→0	0·9→1·6	Up to 1·8
Friesian	5·7→0	1·1→1·8	Up to 2·3
South Devon	5·7→0	1·4→2·0	Up to 3·2

Remarks

Concentrates and hay increased according to appetite.
Milk substitutes slowly reduced as weaning approaches.

The concentrate feed can be in the form of purchased proprietary calf rearing pencils or a suitable mixture can be made up as follows:—

Rolled oats or barley	60 per cent
Linseed cake or beans	30 per cent
White fishmeal	10 per cent

Plus ½-1 per cent of synthetic vitamin A and D supplement

Where separated milk is available, 4·5 litres of this, supplemented with concentrated vitamins A and D, plus 0·45 kg of cereals (oats, barley or maize) fed dry, can be used instead of 4·5 litres of milk substitute.

The Early Weaning System

The First Five Weeks

1-4 *days* colostrum fed as above

5-10 *days*

	Whole Milk litres	*Milk Equivalent* litres
Jersey	1·14	1·14
Ayrshire and Guernsey	1·42	1·42
Dairy Shorthorn and Red Poll	1·70	1·70
Friesian	1·99	1·99
South Devon	2·27	2·27

11-28 *days*

	Milk Equivalent litres
Jersey	2·84
Ayrshire and Guernsey	3·41
Dairy Shorthorn and Red Poll	3·98
Friesian	4·55
South Devon	5·11

29-35 *days*

Reduce milk equivalent by appoximately 0·60 litre per day until calf is weaned on 35th day.

Remarks

Early weaning concentrate mixture (see later) with *good* hay and water available ad lib from the fifth day.

From Sixth to Twelfth Week

Early weaning concentrate mixture plus good hay and water ad lib continues to be fed until consumption reaches the following levels:—

Jerseys	1·8 kg	,,	,,
Ayrshire and Guernseys	2·0 kg	,,	,,
Dairy Shorthorns and Red Polls	2·3 kg	,,	,,
Friesians	2·5 kg	,,	,,
South Devon	2·7 kg	,,	,,

Remarks

Proprietary early weaning pellets can be purchased or an early weaning mixture made up as follows:

Flaked maize	50 per cent
Rolled oats or barley	20 per cent
Linseed cake	10 per cent
Dried skimmed milk	10 per cent
White fishmeal	10 per cent

Plus ½-1 per cent of synthetic vitamin A & D supplement

Hay fed must be of the best quality available.

After the 12-14th week, the early weaning concentrates can be replaced by a concentrate mixture as fed under the 'traditional' system, or proprietary calf rearing pencils can be used.

The early weaning system, by reducing the period of bucket feeding to five weeks, saves labour. It also saves accommodation by reducing the need for individual penning of the calves (see page 97) to five rather than twelve weeks. Once weaned calves can be penned communally according to age.

By early consumption of dry foods, rearing costs are also reduced and early development of the rumen is encouraged. But it should be clearly understood that success in early weaning does depend on using good hay to encourage rumination, and using concentrates of the highest feeding value to ensure an adequate nutrient intake by the calves.

To further reduce rearing costs, spring-born calves (i.e. March-May) can be reared outside on the early weaning system, shelter being cheaply provided by, say, a straw bale hut.

A clean pasture—a new ley not previously grazed by cattle—should be chosen and a rotational paddock system of grazing adopted allowing a total area of 1350 m² per calf. Grazing can then reduce hay and concentrate consumption to not more than 1·36 kg concentrates and 1·36 kg hay per calf per day.

Beef Calves

Calves being reared for beef production, for sale either as beeflings at 18-20 months old or for baby beef at 12-14 months old, can be successfully reared on the early weaning system. But the plane of nutrition must be at the level suggested for South Devon heifer calves on page 91.

Thus, by 12 weeks of age, a beef calf will have consumed, say 18 litres of colostrum, 109 litres of milk equivalent and 140 kg of early weaning concentrates. This compares with the 18 litres of colostrum, 82 litres of milk equivalent and 102 kg of early weaning concentrates that would be consumed by a dairy calf (e.g. Ayrshire) in the same period.

VEAL PRODUCTION

Present-day demand for veal is for calves of 110-140 kg liveweight, well finished to give a carcase of 66-84 kg (60 per cent killing out percentage) over a period of 10-12 week's feeding.

This requires rapid growth—over 0·9 kg/day—with no check, otherwise calves will not give the desired depth of flesh or finish (rather than actual carcase weight) which is required in the trade. Calves for first quality meat should be chosen from breeds with this capacity for high liveweight gain.

The South Devon and Friesian are excellent, as are beef × dairy calves such as the Hereford × Dairy Shorthorn, or Beef Shorthorn × Ayrshire. Other calves can be reared for veal but are unlikely to give first-quality carcases.

House in a well ventilated but draught-proof building, insulated with a low ceiling to give an even temperature of 15-18° C. Individual pens (1·5 × 0·6m) are recommended with a feeding passage—litter can be sawdust or peat moss, straw is less suitable. Slats are probably best—76 mm wide spaced 25 mm apart, and clear of the floor on 150 × 750 mm bearers to facilitate drainage.

Calves can be housed in pens, say 3 × 3 m as in a typical loose box, being tethered with leather straps to the wall. But under this system it is more difficult to inspect the dung from each calf—an important aid to controlling feed intake (see later).

The calf house need not be in darkness, but should be kept

closed to prevent disturbance of the calves except at feeding times. An air extraction fan may be needed if ventilation is poor.

A food-mixing room next to the calf house, with a hot water boiler and a wash-up trough for cleansing and sterilising the feeding buckets, is a great advantage.

Calves can be multiple suckled for veal production, provided the milk yield of the nurse cow is adequate, but feeding for veal today is largely on the basis of using very high-energy milk equivalent feeds. These are now available in proprietary form with minerals and vitamins added—the mineral content of the farm water supply should be ascertained by analysis, as very hard water may upset the mineral balance in the feed.

A guide to the feeding of a veal calf on a veal milk equivalent supplied as a dry meal is as follows:

Guide to Veal Production

(Calves of 36 kg liveweight at birth)

Age (days)	*Milk Equivalent per day* kg	*Water per day* litres	*Feeds per day*
0–4	Colostrum (essential)		3 (or suckling dam)
7	0·5	3·5	3
14	0·9	5·7	2
21	1·4	8·0	2
42	1·8	9·0	2
70 onwards	2·3	11·5	2

No other feed is given

A second method of veal production can be used where, after three weeks of age, dry food is offered ad lib as a supplement to the milk equivalent diet, which is fed at a level of about 1·5 kg per day from this stage onwards. Suitable dry-food veal supplements are available in proprietary form.

The above remarks are intended as a guide only, but note how rapid is the increase of food intake in the early stages. The important rule is to feed to capacity but not to over-feed—each calf should be ready for its next meal.

If appetite is below normal, reduce the quantity of milk equivalent in the next meal by half, but not the water. Some feeders substitute a raw egg (an old remedy), feeding the shell as well, in place of the normal feed. Antibiotics can also be fed to reduce the risk of bacterial scour (see page 96).

Conscientious stockmanship is essential and one of the first signs of trouble is a failing appetite—the temperature of any calf not feeding well should be taken per rectum (see page 116).

Some calves at about 6-8 weeks of age may show loss of appetite due to iron deficiency. This can be overcome by giving one teaspoonful of ferrous sulphate in the milk feed.

The profit margin in veal production is too low to permit of losses through death, or lack of thrift in the calves. It should not, therefore, be regarded as a stop-gap enterprise, but rather as one in which the highest skill in observation, punctuality in feeding, and scrupulous attention to hygiene is required on the part of the stockman.

Some Calf-Rearing Problems

Calf Scours. Profuse diarrhoea or scouring in young calves is either due to a severe digestive disturbance—usually as a result of over-feeding—or to bacterial infections, often of a very mixed character, in the digestive tract.

These conditions may be found together in that bought-in calves, having experienced a period of starvation, may consume a large initial feed and, not being "protected" from the bacteria present on the new farm, may develop scours and die within two or three days.

On the other hand, calves born on the farm are afforded "protection" by the colostrum from their own mothers. It follows therefore that newly-introduced calves need very careful nursing for the first week at least.

Purchased calves arriving hungry at the rearing farm should be given a first feed of two tablespoonfuls of glucose in a quart of warm water. This will alleviate hunger whilst keeping up the energy intake and reduce the risk of nutritional scour.

To combat possible bacterial infection, the calves should be inoculated with *E. coli* serum.

If a calf, or calves, appears sickly on arrival, take its temperature and if over 39° C call in veterinary advice at once. Your veterinary surgeon can prescribe suitable antibiotic treatment, after examining a swab of the faeces if in doubt as to the type of bacterial infection involved.

Infectious scour—manifested by an evil-smelling white diarrhoea—may be endemic in some buildings in which case calves are infected through the open navel at birth. Dressing the navel (see page 120) immediately after birth is recommended. All calving accommodation should receive a very thorough annual spring clean and disinfection.

Calves born outdoors are much less liable to this disease.

Calf Pneumonia. A further serious calf ailment is calf pneumonia, often proving highly infectious and fatal.

This is most certainly a job for the veterinary surgeon, but is less likely to be a source of trouble if calves are kept warm, dry, and free from draughts when young.

Solid partition walls to calf pens are preferred to tubular

DISINFECTION AGAINST CALF DISEASES

1. *Scrape off and remove all litter and dung.*

2. *Follow by scrubbing with warm 4 per cent washing soda (a double handful of soda in 14 litres of hot water).*

3. *Soak the floor and lower part of the walls in an approved disinfectant for 24 hours and later wash out with water.*

divisions and a calf house should not be too lofty. A false ceiling of straw laid on wire netting or wooden slats should be provided in mid-winter if such is the case.

To the observant herdsman, loss of appetite in a young calf is a most important warning symptom. If a calf fails to take its normal feed with relish, look for symptoms of scouring or for a high temperature—over 39° C—indicative of bacterial infection.

The importance of strict hygiene in pail-feeding of calves has already been emphasised; it is this lack of attention to detail which leads many farmers to favour single or multiple suckling as against pail-feeding in reducing calf mortality.

Cross-suckling. The urge to suck is very strong in calves. On bucket feeding they will, unless housed individually, tend to cross-suckle or to suck the ears of other calves.

If calves must be housed in batches, tie them up with leather straps or use individual feeding yokes so that they can be fed individually and only released after their lips are dry.

Hanging a piece of chain in the pen may also prove effective in preventing cross-suckling which may become a most serious vice later in life, particularly in the Channel Island breeds.

Pot-bellied Calves. This condition is due to a slowing down of skeletal growth, in many cases due to too early weaning off milk or milk substitute foods.

To some extent a well-developed barrel is entirely advantageous in calves in that their digestive capacity and ability to deal with long fodder such as hay and silage is increased. But this ideal should not be made the excuse for providing a diet of low nutritive value when under-sized pot-bellied calves are sure to result.

Occasionally one meets calves that, for no apparent reason, fail to thrive—probably due to some physiological defect. I remember one calf which regularly became "blown" after each feed of milk; in such cases early slaughter is advised.

Abnormal Calves. Occasionally calves are born (often failing to survive) with hereditary defects. Examples are the well-known bulldog calves or calves with dropsy, but an abnormality less easy to identify in calfhood is the heifer calf born twin to a bull calf which in later life fails to breed.

Such a free-martin heifer suffers an arrestment of sexual development before birth.

Not all heifer calves twins to bull calves are free-martins; but they are so in every case where the two afterbirths are joined together. If the afterbirths are quite separate the heifer will breed even though twin to a bull, but such cases occur to an extent of less than 1 per cent of all cases of twins of opposite sex.

Where twins are of the same sex no fertility problem need be anticipated.

Unless, therefore, there is a very special reason for taking the risk, heifers born twins to bulls are not reared for breeding purposes.

Another condition which should be looked for in calves is an under- or over-shot jaw—this may make little difference to a calf's ability to suckle or to drink but will certainly affect its ability to graze properly later in life. Such a condition is regarded as hereditary and could result in a bull siring defective stock.

On-The-Farm Operations

Now here are some of the operations that herdsmen may have to perform on calves.

Disbudding or Dehorning

This is particularly recommended where the yarding of cattle is being practised. There are three main methods, all involving the preliminary clipping of the hair around each horn bud:

(a) Use of caustic potash. First, smear a little vaseline or lard around each horn bud. Then moisten a caustic potash stick and rub the moistened end onto the horn bud and continue to rub until the horny layer has been removed and blood begins to appear.

Do this within seven days of birth, but check the efficiency of the treatment a week later. Repeat if any horny tissue still remains.

It is advisable to wear rubber gloves when doing this job.

(b) Use of a hot iron. Electrically-heated and thermostatically-controlled hot-iron calf dehorning outfits are available. The hot iron sears the horn bud and can be effectively used up to eight weeks after birth. This must be done under an anaesthetic.

Under the Protection of Animals (Anaesthetics) Act of 1964, it is illegal to disbud a calf over seven days old except under an anaesthetic.

Earmarking and Removal of Extra Teats

When calves are earmarked for registration purposes it is a good plan to examine them for any extra teats on the udder. At this age, their removal can be easily effected with a pair of sharp, curved surgical scissors, without the risk of damaging the udder tissues.

Tattooing itself should be done carefully. Try to avoid tattooing across the ear veins, thus avoiding septic ears or indistinct tattoo marks owing to incomplete closure of the forceps. For Friesians a red marking ink shows up best. Ear-tags are now acceptable under the Tuberculosis Order 1964, which requires all cattle to be individually identified.

Castration

Where bull calves are reared for beef, castration usually takes place at from two to five months of age. The later in life the operation is performed the more masculine the steer develops in appearance. This can be an advantage when selling cattle as stores.

A method of castration much favoured is the Burdizzo Bloodless Castrator. Young calves can be castrated by this means while standing, but those over, say, three months old are best cast and castrated lying down.

The operator's assistant should sit on the calf's head and keep the uppermost hind leg of the calf pulled well forward. Before closing the jaws of the castrator make sure that the spermatic cord is between them, and make two independent closures for each cord.

Castration with a knife is very effective and probably more reliable than the bloodless castrator but is a job for the veterinary surgeon.

The Animal Protection Act requires the use of an anaesthetic if the calf is over 3 months old and prohibits the use of rubber rings after 7 days of age. The rubber ring method involves the use of a special pair of pliers to place the ring over the scrotum into which the testicles have been safely manipulated.

Vaccination

On some farms vaccination against 'Black Leg' should be carried out before the calves go to grass (see page 130).

BRUCELLOSIS ERADICATION

The programme to eradicate brucellosis from cattle in Great Britain is now well advanced through participation by herd owners in the voluntary eradication schemes coupled with the introduction of compulsory eradication on an area by area basis. By the end of 1977 a large part of England, together with the whole of Scotland and all but a small part of Wales, was subject to compulsory measures. Compulsory eradication will apply to the whole of Great Britain by 1st November, 1979.

When, for practical purposes, the disease can be regarded as non-existent in an area, the area is declared an attested area. By the latter part of 1977 parts of England and large parts of Scotland and Wales had been declared as attested areas.

The Brucellosis Incentives Scheme

This voluntary scheme is intended to encourage owners of clean, or relatively clean herds, to seek voluntary accreditation in advance of the introduction of compulsory eradication in their areas. The incentives take the form of a payment on milk sales from accredited dairy herds and a headage payment for each eligible animal in other accredited herds. The rates of incentive payment are 0·176p per litre for milk sold from a dairy herd through a Milk Marketing Board and an annual payment of £5 for each qualifying animal in other herds. These payments are made for a period of five years from the date an eligible accredited herd is entered on the British Register of Brucellosis Accredited Herds.

To qualify for registration a herd must have reached accredited status by passing a series of qualifying blood tests or have been formed by the transfer of animals of accredited status to previously approved premises. No compensation is payable for reactors required to be slaughtered, nor in the case of herds outside eradication areas, for dangerous contact animals. During the five-year incentive-earning period the owner is recommended to set aside part of his incentive payments to provide insurance cover for the cost of replacing reactors arising from re-infection.

After expiry of the incentive earning period, the Ministry/Department is prepared to arrange the slaughter of reactors and to pay compensation at 75 per cent of market value subject to a maximum payment per animal, details of which may be obtained from Divisional and Animal Health offices of the Ministry/Department. The Incentives Scheme is closed to new applicants from the start of compulsory eradication in an area.

Brucellosis (Accredited Herds) Scheme

Herd owners remaining in this scheme (which was replaced by the Incentives Scheme) retain the existing benefits until 31st March, 1980, including full compensation for reactors subject to a maximum payment per animal. Thereafter these owners will be eligible for the same terms as those available to owners of herds in the Incentives Scheme upon expiry of the incentive earning period.

Compulsory Eradication

When compulsory eradication measures are applied in an area, all herds not participating in the voluntary schemes are tested with the object of identifying and removing any reactors to enable these herds to reach accredited status. Movement controls are applied and vaccination with 45/20 vaccine is prohibited except under licence. Reactors and dangerous contacts are removed for slaughter by the Ministry/Department with payment of compensation. In herds which are not yet accredited the compensation for reactors is 100 per cent of market value and in the case of accredited herds 75 per cent of market value subject in each case to a maximum payment per animal. Full market value without limit is paid for contact animals required to be slaughtered.

Attested Areas

When the incidence of brucellosis in an eradication area has been reduced to a very low level consideration is given for the area to be declared an attested area. In an attested area there is some relaxation of the restrictions that apply to the movement of animals within the area, but other restrictions on movements into the attested area are imposed with the aim of preventing the introduction of disease from outside. No vaccination with either S19 or 45/20 vaccine is permitted.

Calf Vaccination

The Ministry/Department provides, as a protection against brucellosis a free service for the vaccination with S19 vaccine, of female calves intended for breeding. To be eligible for this service calves must be between the 91st and 180th day of life. This service is, however, not available in attested areas where vaccination with S19 is prohibited.

British Register of Brucellosis Accredited Herds

This Register is the only official list of accredited herds. Herds are automatically entered on the Register if they secure and retain accreditation through either the voluntary Brucellosis Incentives Scheme; or the earlier (Brucellosis Accredited Herds) Scheme; or compulsory eradication arrangements.

British Register of "Officially Brucellosis-Free" Herds

The EEC Directive which governs intra-Community trade in bovine animals prescribe detailed tests and procedures for the classification of herds as "officially brucellosis-free". Although there is provision for the continuing use of UK procedures for designating herds as "brucellosis-free" (i.e. accredited herds) the Directive requirements must be met to achieve the classification of "officially brucellosis-free". The Directive also lays down that non-slaughter stock for export to other Member States (i.e. breeding stock or stores) must have originated from a herd "officially brucellosis-free" unless the importing Member State has agreed to allow cattle to come from herds classified as brucellosis-free.

To assist herd owners wishing to trade with other EEC countries in exporting breeding and production cattle, a British Register of Officially Brucellosis-Free Herds has been set up. To qualify for entry on the register a herd must have previously achieved accredited status, contain no females vaccinated against brucellosis during the previous three years, have been free of clinical signs of brucellosis for six months and all the animals in the herd who were 12 months of age (including steers) must have passed blood tests at the required EEC standards.

CHAPTER VII

FEEDING AFTER WEANING

AFTER weaning there follows a period of some 12-18 months duration during which bodily development continues—but at different levels of nutrition for beef and dairy stock.

In general, the level of feeding is higher for cattle being reared for beef than for cattle being reared for dairy purposes. The aim with beef animals should be to prevent the "store period" as it is called being so severe that the conformation of the cattle suffers.

Under conditions of poor feeding the head, horns, tail and legs continue to develop, whereas spring of rib and width of loin and development of hind quarters are slowed up. This gives the long-legged, shallow-ribbed, and large-headed conformation one often sees in store cattle which have been badly wintered. Since the most valuable part of the carcase is the loin and hindquarters such conformation is against the interests of quality meat production.

Keeping Rearing Costs Low

However, the store cattle rearer has also to endeavour to keep rearing costs as low as possible and it is to that end that a store period (generally in winter) is justified—provided the level of nutrition does not fall too low.

During the store period the opportunity occurs to use the poorer-quality fodders such as oat straw and second-quality hay, with some supplementation by silage or roots. Then, usually, satisfactory progress can be achieved without recourse to more expensive feeding of cereals or oil cakes.

There are two groups of cattle that are exceptions to the above generalisations: baby beef animals and bull calves being reared for stud purposes.

In both these cases the level of feeding should be maintained throughout the rearing period so that no store period as such is experienced. This allows the full inherited growth rate to be achieved, giving well-grown bulls (capable of earlier mating) or well-finished, early-maturing baby beeves.

Level of Feeding

Allowing for breed differences (i.e. rate of growth, which is related to the mature liveweight of the breed as well as to early maturity), the differences in feeding levels between cattle reared for different purposes can be illustrated in the following table:

High Level of Feeding

(Applicable to bulls for stud and baby beeves, generally reared by single suckling.)

Average daily gain—0·8-1 kg per head

Typical liveweights	8 months—250 kg
	12 months—350 kg
	14 months—430 kg

Typical Winter Rations	*At* 8 *months*	*At* 12 *months*	*At* 14 *months*
	kg	kg	kg
Hay (good)	2·7	4·5	5·4
Roots	12·7	12·7	12·7
or			
Silage	9·5	9·5	9·5
Concentrates	2·7–3·6	3·6–4·5	4·5–5·4
		No store period (12–14 months)	

Medium Level of Feeding

(Applicable to cattle reared for first-quality beef (and to heifers of the British Friesian, Dairy Shorthorn, South Devon and Red Poll breeds for early calving) generally reared by multiple suckling.)

Average daily gain—0·6–0·7 kg per head

Typical liveweights	*Beef cattle*	*Dairy cattle*
	kg	kg
8 months	200	180
16 months	350	300
24 months	480	430

Typical Winter Rations	*At* 8 *months*	*At* 16 *months*	*At* 24 *months*
	kg	kg	kg
Hay (medium)	2·7	3·6	5·4
Straw (oat)	—	3·6 (to appetite)	3·6 (to appetite)
Roots	12·7	15·9	19·1
or			
Silage	8·2	10·0	12·7
Concentrates	1·8	None (Dairy cattle) 0·9 (Beef cattle)	None (Dairy cattle) 0·9 (Beef cattle)
		Moderate store period	

Low Level of Feeding

(Applicable to dairy heifers generally reared by pail feeding.)

Average daily gain—0·45-0·60 kg per head (dependent on breed).
Typical liveweights

8 months—140 kg
16 months—250 kg
24 months—370 kg

Typical Winter Rations	*At* 8 *Months*	*At* 16 *months*	*At* 24 *months*
	kg	kg	kg
Hay (medium)	2·3	2·7	4·5
Straw (oat)	—	2·7 (to appetite)	4·5 (to appetite)
Roots	6·3	9·5	12·7
or			
Silage	4·5	6·8	9·5
Concentrates	1·4	None	None
		Moderate store period	

The concentrate mixtures would normally be compiled as given on pages 78 and 79.

In principle, therefore, the higher planes of nutrition for beef production or for inducing rapid bodily development are associated with higher levels of concentrate feeding and lower levels of roughage foods.

This adds to the expense of winter feeding, and with dairy heifers economy is achieved by feeding hay, straw, roots and silage to appetite. Given reasonable quality in these home-grown foods, concentrate feeding should not be necessary after 12 months of age, except for bull calves and baby beeves.

The control of the plane or level of nutrition is much more positive in winter than in summer feeding. But, even in summer, by regulating the rate of stocking and by the selection of pastures some degree of control can be exercised on rate of liveweight gain.

Thus dairy farmers commonly use young cattle as "followers" to the dairy herd, and their function then is to make fuller use of the grazing by "cleaning up" after the most nutritious first grazing has been eaten by the more productive in-milk stock. (This should not be done with young calves.)

Similarly, young beef stock can be used for grazing second-rate and third-rate pastures not capable of fattening cattle.

"Worth an Extra Feed"

Young cattle are generally in-wintered for at least their first winter, i.e. when less than 12 months old by the autumn. There is a very true old saying "that a dry back and a dry bed are worth an extra feed".

In-wintering saves food and saves poaching of pastures, but where sheltered dry-lying land is available out-wintering can be carried out without detriment to the cattle or the land and will save labour in feeding, litter and manure handling.

On arable farms, however, the loss of dung is a disadvantage—unless outwintering is possible on land due for the plough.

Yard accommodation for young cattle need not be elaborate. Mainly open yards with some overhead shelter are probably best in that the cattle tend to grow a good coat of hair and do better when they go to grass in the spring than when completely housed.

Around 5·6 m^2 of yard space is desirable per beast; rather less if dehorned or polled. There must be ample trough room—up to 1·2 m per head.

Concentrates, roots and silage are usually fed in troughs or bins, and long fodder in racks. Facilities to enable these to be filled from the outside of the yards are great labour-savers.

Yarding in Groups

Young cattle should be yarded in groups of comparable size to avoid bullying; otherwise the poorest individuals get progressively poorer and may eventually need to be taken out of the yard and individually fed.

The alternative to yarding is to tie up young cattle in sheds. This saves space but adds to labour in feeding and cleaning out and may result in over-grown feet and swollen joints owing to lack of exercise.

A close watch should be kept for outbreaks of lice, mange or ringworm and prompt remedial measures taken. In the handling of cattle in yards a cattle crush (either built in permanently on a site near the yards or a movable cattle crush as now available) is highly desirable.

Turning Out to Grass

When young cattle are turned out in spring they often suffer a considerable check to progress and lose weight, which requires extra time to restore.

The cause is not only the change from being indoors to being more exposed to weather conditions (and weather in May can be cold and wet) but is coupled with the change from a diet of mainly dry foods to a very laxative diet of grass.

This change should be cushioned by making the turning-out process a gradual one, housing the cattle in the yard at nights for a week or ten days and giving them a feed of 2-3 kg hay per head daily before turning out to grass.

Some authorities have recommended the feeding of a costive type of concentrate such as undecorticated cotton cake but this today would prove too expensive even at only 1·4 kg per head daily.

Where no yarding accommodation is available choose a sheltered field for turning out and turn out before the grass is making its full flush of growth so that the grass grows to the cattle. Under such circumstances the cattle will eat some hay if offered to them, with every advantage.

It is not usual to turn calves to grass until they are around six months old, but it can be done sooner provided the above feeding precautions are observed and the pastures are clean, i.e. they have not been grazed by cattle for at least a year, thus reducing the serious risk of husk infection.

Should coughing be observed in young calves, particularly when driven, house them immediately and feed generously; husk, which is referred to again in the next chapter, is a most debilitating and even fatal disease in young calves.

Experiments in New Zealand with identical twins have shown that calves rotationally grazed on new leys outweigh the opposite twins grazed on low-quality pasture by over 50 kg at ten months old. The old practice of always using a paddock convenient to the buildings as a permanent calf paddock, a common practice on many farms in Britain, cannot be too strongly condemned.

Better Bulk Converters

In the rearing of calves for beef a series of very carefully conducted experiments by Mansfield and Brookes at Cambridge has shown the economic justification for rearing calves on a high plane of feeding up to at least eight months of age.

In other words a liveweight gain of over 0·7 kg per calf per day should be aimed at. Calves reared in this way are then better able to utilise cheaper bulk foods earlier in life, and economy in concentrate feeding is achieved.

Underfeeding of calves so that their high growth potential is not realised means a slow-growing, stunted calf, which only reaches slaughtering weight after a longer period of feeding.

This is not so vital in the economy of rearing dairy heifers where early maturity is less necessary—unless calving is advanced to an earlier age. In fact some dairymen hold the view that heifers should not be heavily fed, but this should not be made an excuse for failing to give all calves up to weaning age a good start in life.

By adequate feeding in early calfhood better utilisation of food is achieved than by stunting the calf and resorting to heavier (and, therefore, more expensive) feeding later.

When to Serve

The level of nutrition is also important in promoting normal heat periods in heifers due for bulling. Such heifers must be well fed, otherwise—particularly in winter—they may fail to come on heat even when running with the bull, or, if to be inseminated, may fail to show heat symptoms or be on heat for only a short time. During the long winter nights they may therefore escape detection.

Well-grown heifers can be served younger than less well-developed heifers. So size rather than age should determine whether a heifer is suitable for service or not, bearing in mind that some breeds, notably the Channel Island, are sexually earlier maturing than other breeds.

The economic aspect of early calving, in that rearing costs are repaid sooner by early calving, supports the practice—provided growth has been normal for the breed.

The following liveweights are suggested as normal for the various breeds:—

BEEF BREEDS

	Liveweight kg	*Heart girth* mm
Aberdeen-Angus	360-370	1,640
Hereford, North ... / Devon, Sussex ... / Galloway	420-430	1,700

In the case of dairy breeds the approximate ages of the bulling heifers are added:—

DAIRY BREEDS

	Liveweights of bulling heifer kg	*Normal age at bulling* (months)	*Heart girth* mm
Jersey	229—241	15—18	1,420
Ayrshire and Guernsey ...	292—305	18—21	1,340
Dairy Shorthorn and Red Poll	356—368	21—24	1,630
British Friesians	381—394	21—24	1,660
South Devon	419—432	21—24	1,700

New Feeding Techniques in Beef Production

Reference has already been made to the high plane of feeding associated with the rearing of stud bulls and baby beeves (page 102) and conventional rations have been illustrated.

Within the last decade, work by Dr Preston at the Rowett Institute has evolved systems of *ad lib* feeding on a mainly barley diet with cattle entirely indoors until slaughter—so-called "broiler" beef. Calves can be reared on the conventional early weaning system, being fed *ad lib* from 12 weeks old on a ration such as the following:

Rearing Ration—3-6 months	*kg*
Flaked maize	400
Rolled barley	400
Molassine meal	100
Fishmeal	50
Decorticated groundnut cake	50
Minerals	9

plus vitamins A & D as per instructions supplied

No hay is fed and the calves are best bedded on sawdust or shavings, though experience since has shown that provision of a little hay is beneficial.

By six months old, consumption is up to 6 lb (2.7 kg) of this ration per day. The calves are then gradually introduced to the fattening or finishing ration as shown below:

Fattening Ration—6-12 months	*kg*
Rolled barley	100
Groundnut cake	8
Minerals	1·45

plus vitamins A & D as per instructions supplied

Fresh clean water must always be available and 150 mm of trough space should be allowed in the self-feeding hoppers, with 3·25 m^2 of yard space per head. Covered or partially-covered yards to reduce exposure are best. Ventilation is very important and, above the 1·5 m level, walls are suitably made of vertical 150-200 mm timber boards placed with 13 mm gaps between each board and the next.

The barley must be rolled—barley of 17-18 per cent moisture is recommended. Fine grinding—liable to cause bloat and excessive production of lactic acid in the rumen—must be avoided.

At about 270 kg liveweight the cattle can be implanted behind the ear with 60 mgm of hexoestrol, which will improve growth rate during the subsequent 10-12 week fattening period by 15-20 per cent.

Importance of Conversion Efficiency

Fed on an all-concentrate diet, beef cattle compete directly with pigs and broiler chickens as meat producers, as opposed to traditional beef production which offers a means of cashing

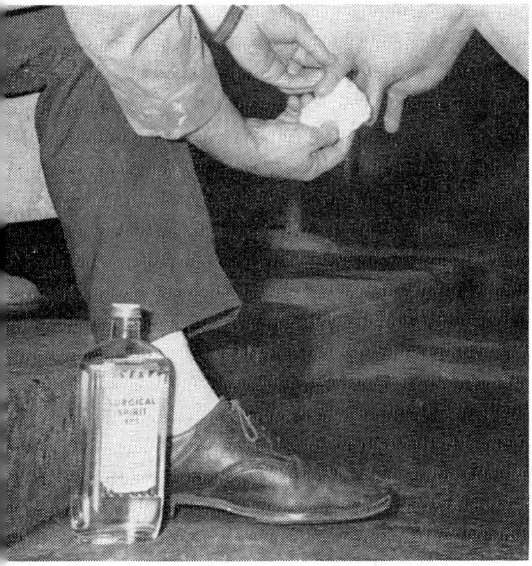

Dry cow management

Teats should be cleaned with surgical spirit immediately after last milking.

If mastitis has been a problem, a precautionary infusion of penicillin should be given.

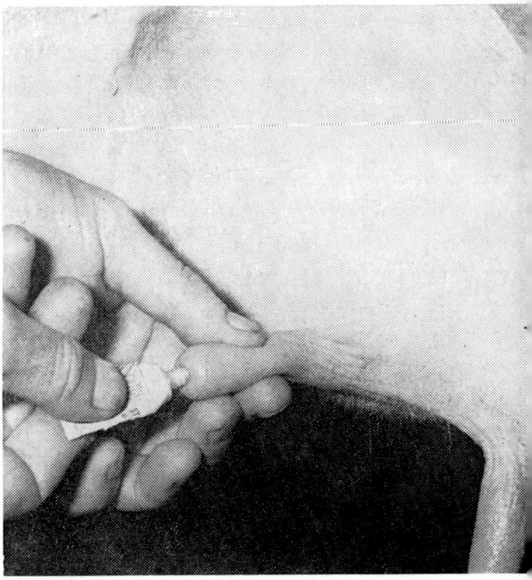

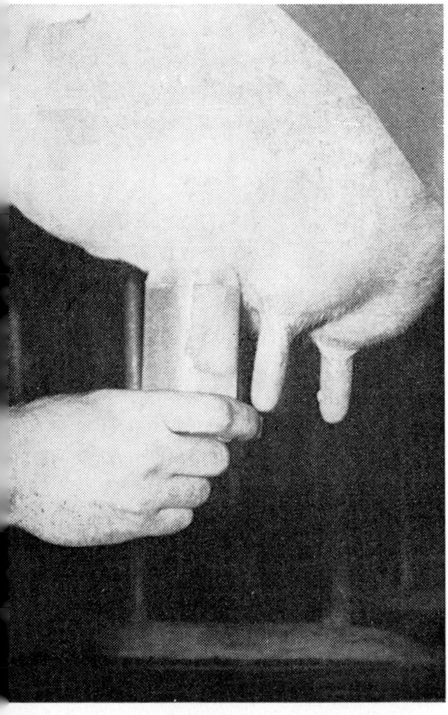

Finally, to prevent bacteria getting into the udder during the dry period, teat orifices should be sealed by dipping each teat in a cupful of iodised collodion.

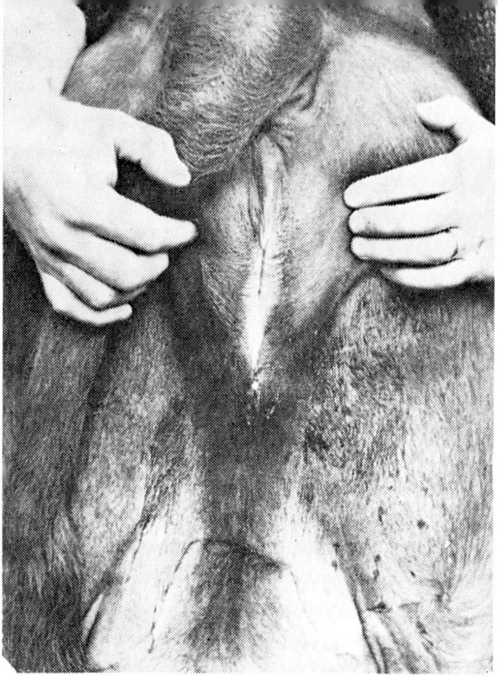

As calving approaches, the vulva becomes grossly enlarged, with the lips parted slightly, and there will be a discharge of mucus from the vagina.

Signs

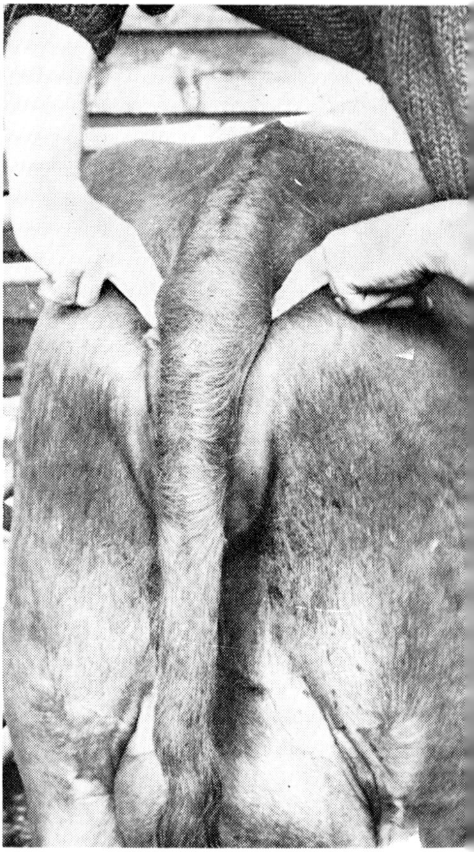

Tailhead muscles slacken and a sunken area appears between each pin bone and root of tail, at first taking a finger but big enough to take a fist just before calving.

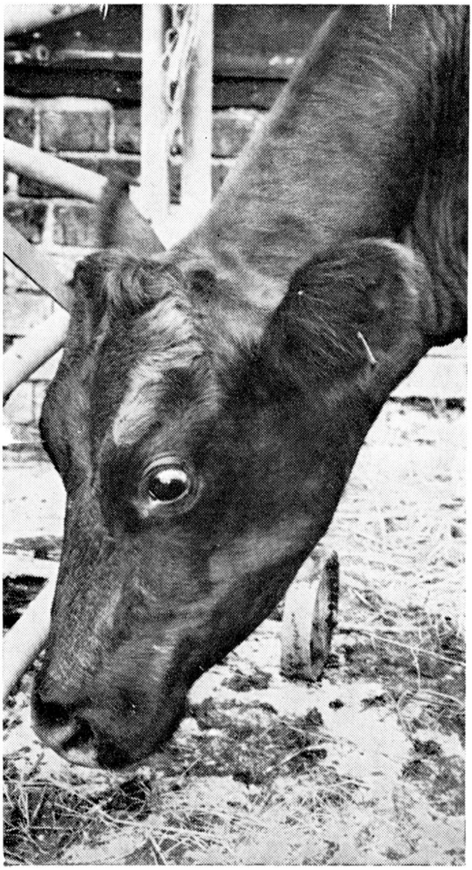

Cow becomes increasingly uneasy, folding her ears back, rolling her eyes, mooing and moving irregularly.

of calving

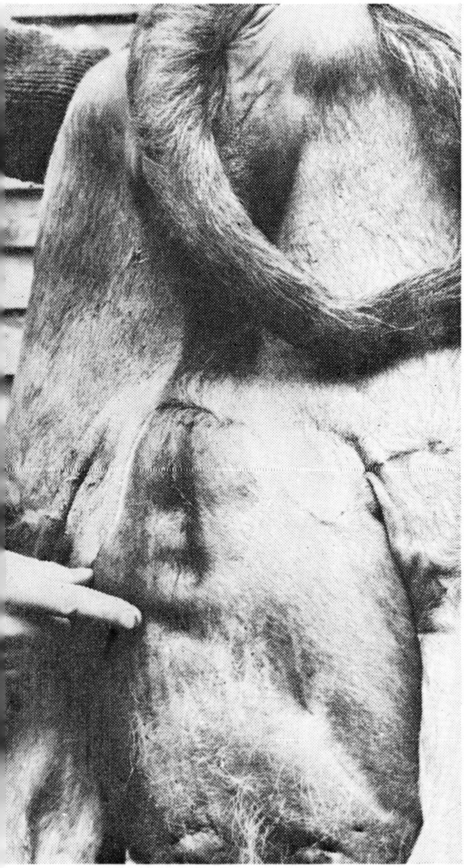

Twelve hours before calving udder is fully distended and pressure with fingers leaves a deep impression.

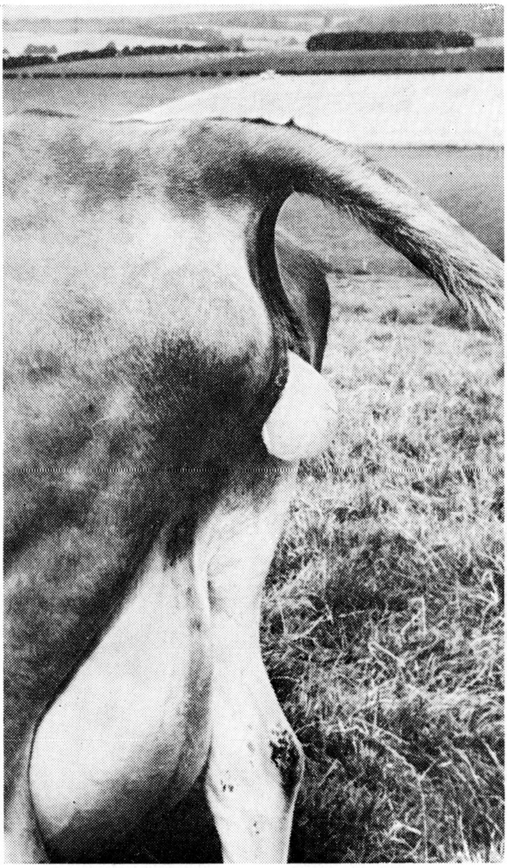

Appearance of water bag indicates calving is imminent, but often bursts soon after appearance and is not always spotted.

Calving without complications

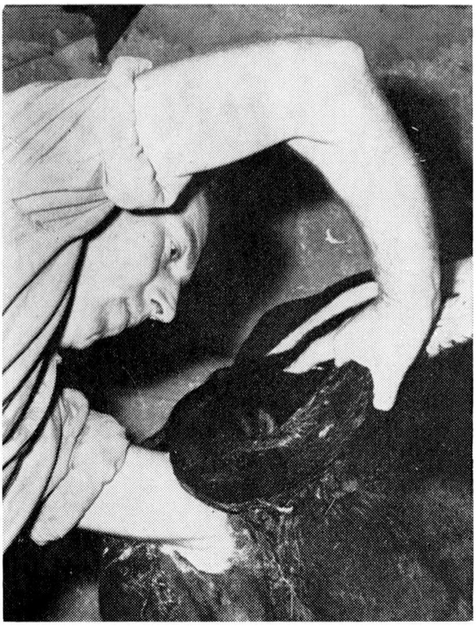

Presentation should be checked 2 hours after the cow starts straining, 3 hours with a heifer—after washing and lubricating the arm with soap and a warm solution of non-irritant antiseptic. It should be possible to feel the head and front feet if presentation is correct.

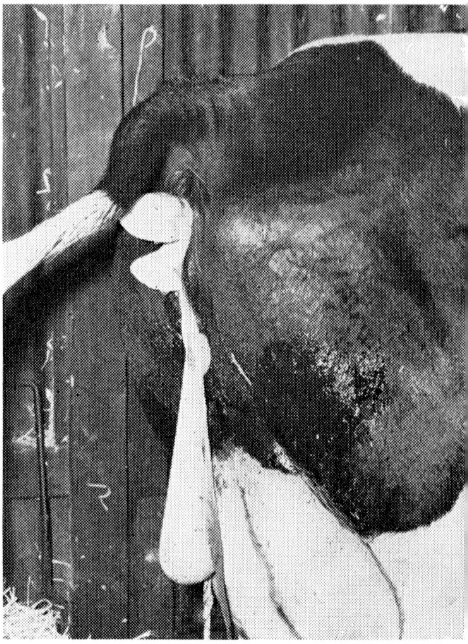

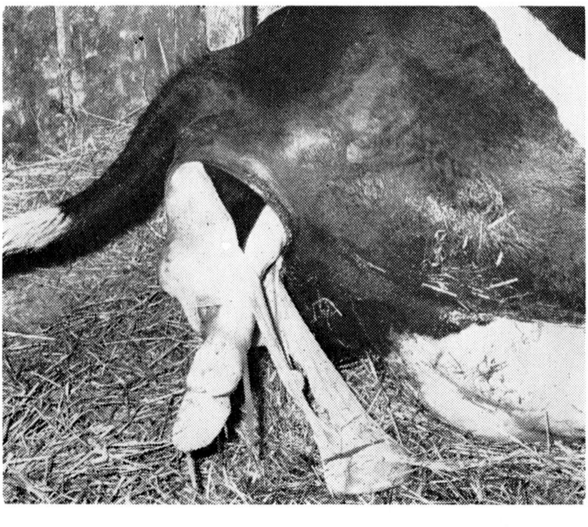

After the water bag has appeared and burst, calf's forelegs should appear within an hour.

During course of labour cow will usually lie down. Forelegs of calf, initially presented sideways, will rotate through 90° to the horizontal position, and will be followed by the head.

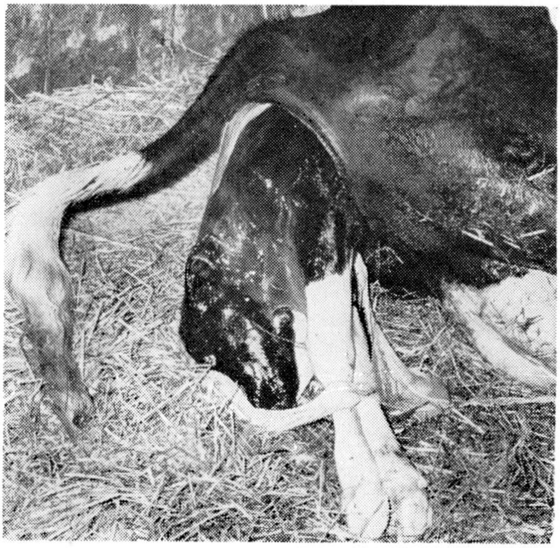

After the head has appeared, the rest of the body should be expelled rapidly.

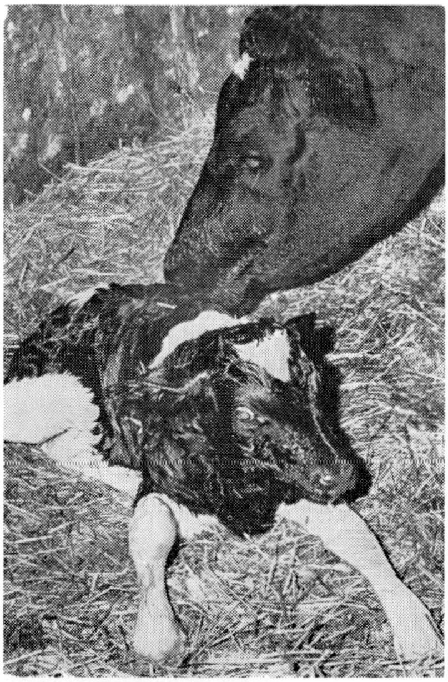

When the calf has been born, the cow should get up and start licking it. This will remove adhering membranes and stimulate the calf.

A strong and healthy calf will stagger to its feet within 10 to 15 minutes and have found the cow's teats within half an hour.

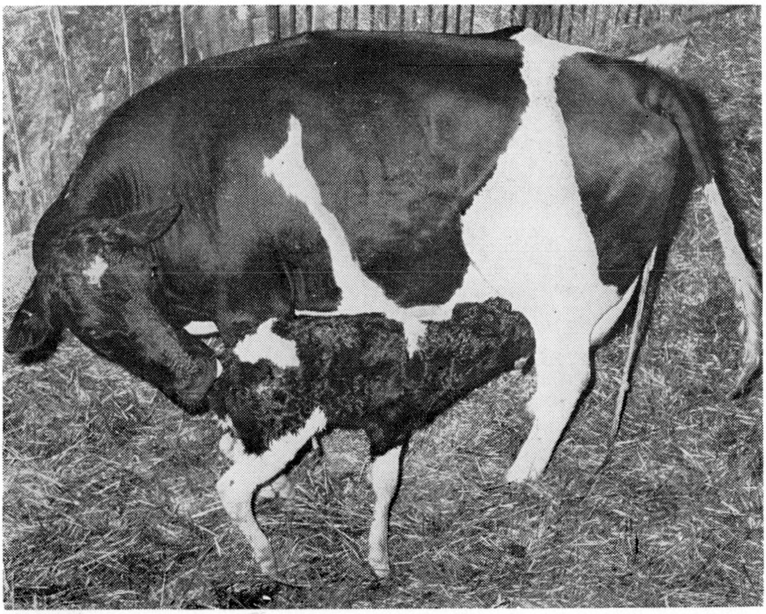

Remove any mucus round nostrils and mouth that may hamper breathing.

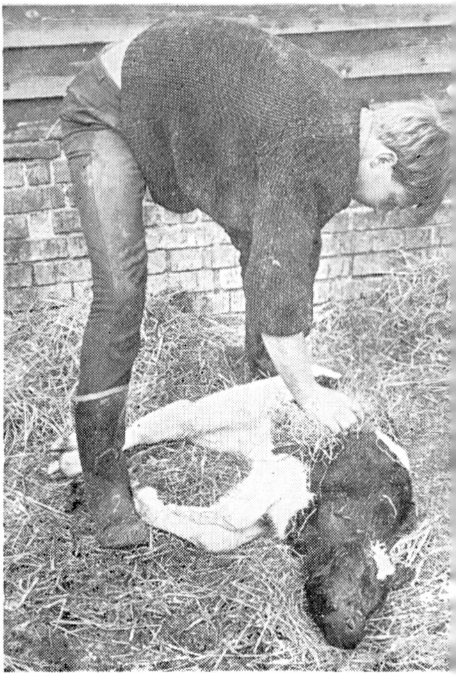

Brisk rub-down with sack or handful of straw will stimulate blood circulation and get calf on its feet.

If not breathing properly try slapping briskly on side of rib cage.

If this is not effective, shock from throwing cold water over calf's head may induce breathing.

Care of the Calf

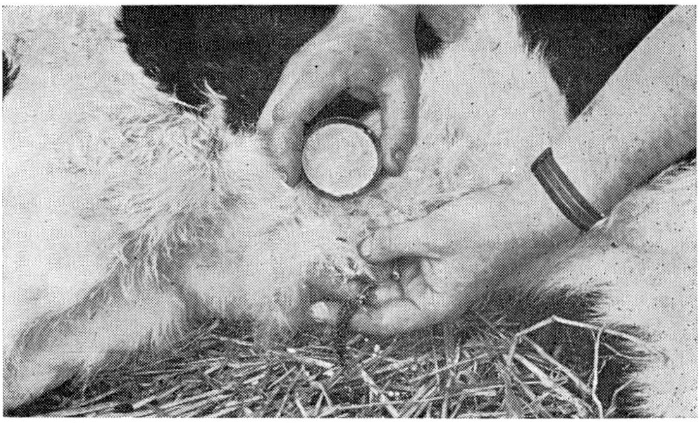

To prevent infection through the navel, apply an antiseptic as soon as possible after birth.

If both steps fail, try artificial respiration by laying calf over on its side and working upper leg briskly round in a circular motion.

The 'kiss of life' is another possibility. Block the mouth and one nostril and blow into the other alternatively with pressing on the ribs to deflate the lungs, repeating the process until natural breathing starts.

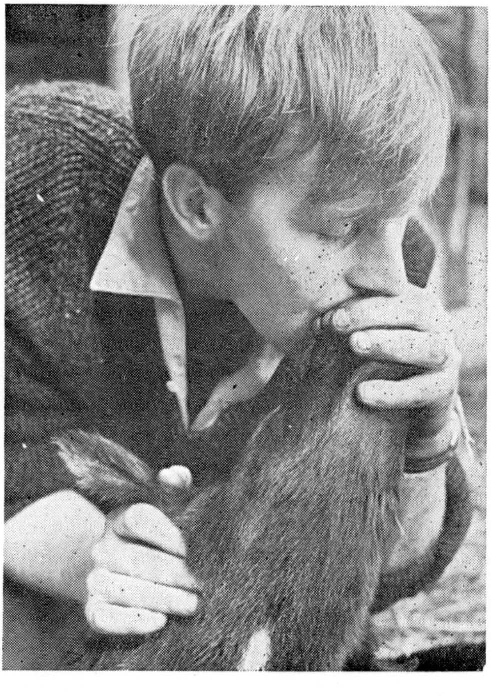

Foot trimming

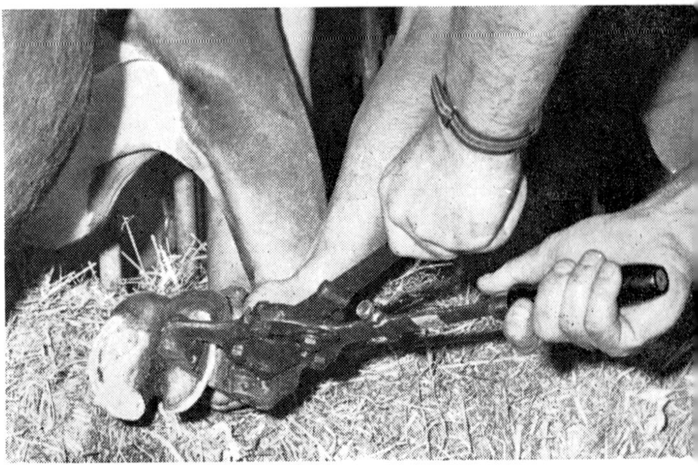

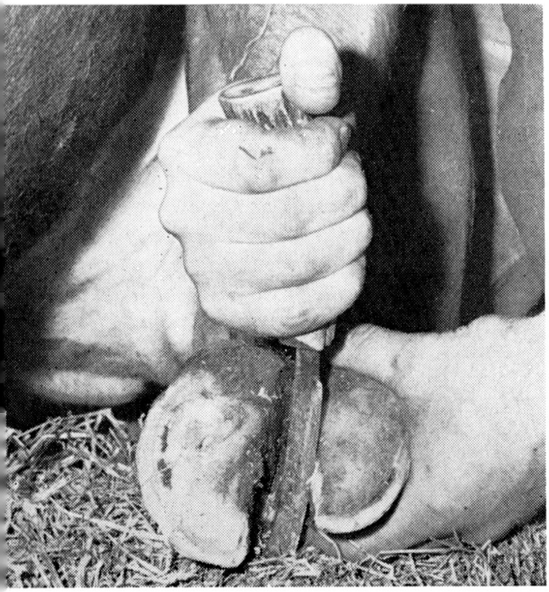

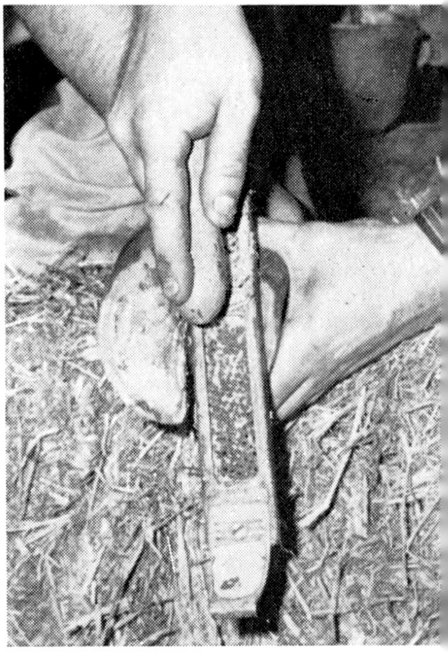

After clipping as much surplus horn off the toes as possible with the feet on the floor, hold up each foot in turn and continue paring overgrown horn from the walls of the claw. Work round each edge two or three times, taking a little at a time to avoid cutting into the quick.

Then use a sharp paring knife to remove surplus horn from inside edge of claws to widen cleft between them.

After roughly paring down the sole with the knife, use a coarse rasp to obtain a clean, smooth and slightly dish-shaped surface and to smooth off round the walls of the hoof.

arable crop by-products (straw, beet tops, pea silage) and grazing in the form of meat.

Given accommodation and ample supplies of cheap barley, this system is as much a "factory" enterprise as pigs or broilers.

As in both of these enterprises, the conversion efficiency of food into meat is fundamentally important. With this type of beef production, conversion efficiency falls off with increasing age and liveweight—above 450 kg liveweight the cost per kg of liveweight gain is too high in relation to present beef prices.

What is required then, is calves not too expensive to buy and capable of a high rate of liveweight gain (well over 0·9 kg/day). To date British Friesian steers have proved very successful.

Further, being fed entirely indoors, the cattle should be marketed during the period February/June, when beef prices are highest, rather than July/November when they will have to compete with grass-fattened beef.

Lastly, cheap barley is obviously basic to the economy of such beef production. A 450 kg baby beef animal reared on this system will consume around 2,000 kg of concentrates of which 1,500 kg would be barley—the produce, say, of 0·4 hectare of barley giving a *net* output of 783 kg of liveweight gain per hectare.

This is much more than could be obtained from the same land under grass. However, the same quantity of barley fed to pigs with a 10 per cent high protein additive would produce eight pork pigs weighing, say, 450 kg, giving a *net* output of 408 kg of liveweight, producing 288 kg of pig meat compared to 191 kg of beef.

Only, therefore, in times of barley surpluses and a shortage of beef, leading to high prices for beef relative to pig meat, is this system likely to have a very wide appeal.

Writing in the spring of 1978, it is clear that the high price of concentrates sets severe limits to the use of all intensive systems of meat (including beef) production for the time being.

CHAPTER VIII

HERD HEALTH AND HOW TO MAINTAIN IT

THE causes of ill-health are many—as are the degrees of its severity—and the standard of health regarded as satisfactory by some farmers will not be the same as by others requiring a higher level of performance or freedom from detectable disease.

Ill-health is, therefore, a very relative term. Nevertheless cattle in any state of ill-health are not fit to give maximum performance, whether as meat or milk producers or as breeding stock.

EARLY SIGNS OF TROUBLE

So a herdsman should be continually alert for signs of ill-health in the very early stages. Any departure from normal behaviour, including the normal sexual behaviour, should be carefully noted.

Cattle in poor health frequently move away from their herd mates if at pasture; indeed loss of appetite is a very early sign of ill-health.

Another is an abnormal posture—standing with a dejected appearance—ears and head held low and back arched; or, when lying down, the head is either stretched out or turned towards the flank.

The coats of cattle in good health are bright and glossy in appearance with evidence of lick marks, and the tails and hindquarters should be free of heavy faecal contamination or discharges from the genital organs, which often cause a "shiny" appearance to the pin bones on each side of the tail root.

Watch for these Symptoms

The general attitude should be one of alertness, and if disturbed the response should be reasonably quick. Cattle that are dull and listless or slow to move should be watched carefully for any signs of lameness, or (in milking cows) for acute mastitis, or (in recently-calved cows) genital infection after calving. Another possibility could be a foreign body in the stomach wall.

Heavy abdominal breathing or irregular movements of the flank or the ribs denote trouble. The nose should be damp and free of mucus.

Pain should not be experienced when the body and limbs are handled, and the skin—handled over the last rib—should feel sleek and velvety between the fingers.

The horns and ears should not feel "cold"; the eyes should be bright and not sunken, and the mucous membrane within the lower eye-lid should be pale pink and not anaemic in appearance.

Finally, cattle should be observed to see if they cud normally (generally within three-quarters of an hour of a meal) and if the dung and urine are normal. Very loose or very firm dung both indicate some derangement of the digestive system owing to unsuitable food or the presence of disease.

Strong-smelling dung or urine, or blood-stained faeces or urine are further abnormalities to be noted. So are abnormalities in milk either in appearance—as will be shown by using a mastitis strip cup—or in smell such as the smell of acetone (pear drops) which indicates acetonaemia.

Three Health Checks

There are three preliminary checks on health which can be made on the farm:

1. THE PULSE, which measures the rate at which the heart is beating. The heart beats can best be felt by gentle pressure on the artery underneath the tail, using the tips of the first three fingers.

Count the number of pulsations in one minute; the normal pulse rate is from 40-60, a pulse rate over 80 is dangerous.

2. RESPIRATION. Count the number of chest movements per minute. The normal breathing rate in cattle is 25-30 per minute.

3. TEMPERATURE. Using a clinical thermometer (see that the mercury is shaken down into the bulb), apply a little vaseline and then insert into the rectum, bulb first. Leave about one inch of stem showing. Withdraw after at least three minutes then take the mercury reading. Normal temperature in cattle ranges from 38-39° C.

That is about as far as the herdsman can be expected to go in the diagnosis of disease. Treatment is the province of the veterinarian, but the value of knowing preliminary symptoms is that the herdsman is better able to decide when to call in professional advice.

However, I have added at the end of this chapter some notes on first-aid and on the more detailed symptoms of the most frequently-met diseases. And before coming to those it may be well to dwell in greater length on what is perhaps the most vital aspect of cow management from a health point of view—the sexual cycle.

Signs of Heat Period

Sexual maturity varies with the breed. It is reached earliest in the Jersey and latest in the hill breeds, e.g., Highland cattle. But females usually exhibit the first signs of "bulling" or oestrus soon after 12 months of age; and under good nutritive conditions normal heat periods occur at 19-23 day intervals.

When on heat cattle are restless, often bellow frequently and attempt to mount other females and will stand to allow other females to mount or ride them. Tell-tale evidence of a cow having been in season is found in the form of disarranged, worn or muddy hair on her rump.

Oestrus or heat lasts for 12-24 hours, longer in summer than in winter. In fact in winter the manifestations of heat may be very brief and pass unnoticed with housed cattle, though oestrus is always accompanied by some swelling of the lips of the vulva and more copious production of mucus.

Females which come on heat at more frequent and irregular intervals than 19-23 days may be suffering from a disease or cystic condition of the ovaries and should be the subject of professional advice and treatment.

Similarly a failure to come on heat after calving may indicate a partial suspension of sexual activity, due to the presence of diseased ovaries or a breakdown of the cycle whereby ovaries return to breeding condition after calving, a matter that can be successfully treated by the veterinary surgeon.

Poor feeding will also cause females to fail to come on heat, as will over-fatness.

Venereal Diseases

The resumption of heat periods after a three to four-month period during which the cow or heifer was thought to be in calf should be regarded with great suspicion. Vibrio foetus, or trichomoniasis, both infections which are venereal diseases causing early abortion—may be responsible, and before any further services females should be subjected to veterinary examination.

With bulls lack of sexual desire or failure to serve may be due to overfatness or to pain associated with actual service where the hind feet are badly overgrown.

Lack of desire may be due to over-use or unfamiliar surroundings; for example, some bulls refuse to work in a service crate. In such cases walking the bull up to the cow and then away from the cow once or twice, without allowing service, will generally stimulate his desire.

As a last resort, turning a bull loose with a cow will generally result in successful service even when he will not serve whilst on a chain or staff.

Infertility in the bull due to the production of sterile or low-vitality semen is impossible to detect without microscopic examination of his semen. Whether a bull is "right for stock" does not therefore depend merely on his ability to serve successfully.

Signs of Pregnancy

A normal pregnancy with cattle is within the range 278-288 days. Bull calves tend to be carried longer than heifers and with twins pregnancy tends to be shorter than normal.

From conception to calving heat periods are absent—though I have known cows stand to the bull during pregnancy, and rare cases have been recorded of a "dual" pregnancy with one calf being born three weeks after another.

The next sign of pregnancy after absence of heat periods is seen about the fifth month after service. Then the udder tissues in a heifer begin to develop more quickly, and with in-milk cows a definite drop in milk yield occurs—often to the order of 20 per cent in one month compared with the normal decline of 10 per cent.

In heifers the udder tissues begin to fill with a sticky fluid, clear at first but later becoming more brown in colour—a very useful confirmatory test of pregnancy.

A veterinary surgeon can confirm pregnancy earlier than the fifth month by manual examination of the genital organs through the wall of the rectum. A further confirmation is the change in disposition of the pregnant female to a more placid nature, and—provided nutrition is adequate—the bodily condition markedly improves.

Feeding the Pregnant Cow

Nutrition during pregnancy is of the greatest importance in ensuring the birth of well-developed, healthy calves and an adequate milk supply in the dam. Ideally, throughout pregnancy there should be a steady gain in liveweight of the dam from the relative lean condition at service to the well-fleshed but not fat condition at calving time.

In dairy cows (which are normally dry for only six to eight weeks before calving) extra concentrate feeding is now a recognised practice to ensure full development of the udder tissues and to provide bodily reserves for the following lactation (see chapter III).

Such a practice is not desirable with pure beef cattle which, under such treatment, would become too fat and may have too much milk for single-calf suckling.

As Calving Approaches

As calving time approaches a cow or heifer exhibits definite symptoms. First, the udder shows a rapidly increasing distension with stiffening of the teats. The appetite may decline slightly, or a marked appetite for certain foods may be shown.

At pasture a cow or heifer near calving seeks a secluded spot or the shelter of a hedge and frequently rises to move round slowly with the tail arched and, when lying, shows labour pains or straining to calve.

Valuable cattle should be kept under close observation. It is a wise practice to house a dairy cow in a loose box (which has been carefully disinfected and provided with plenty of short straw as litter) from the time the udder begins to fill, that is for two to three days prior to calving.

American data suggests that a 0·5-1° C drop in temperature is a sure sign of calving being likely within 12-24 hours—as is the loosening of the muscles around the tail and a sunken area appearing between each pin bone and the tail root. The vulva becomes greatly enlarged and there is a mucus discharge from the vagina. As labour pains and straining become more pronounced the cow generally lies down.

In a normal presentation the water bag containing the clear allantoic fluid is the first to appear. It soon bursts and the cow often rises to lick this fluid—an action which serves to arouse her maternal instincts; in the field she may mother other calves if present.

Within 40 minutes the fore feet of the calf should appear, enclosed in a thicker membrane containing the chorionic fluid which serves as a lubricant to assist the expulsion of the calf.

If birth proceeds normally the calf's head appears soon after, followed by the shoulders and eventually the complete body of the calf.

In such a normal birth the herdsman can give assistance to the mother by pulling gently on the fore feet of the calf in a downward direction towards the cow's hocks, and as soon as the calf's head is clear wipe the nostrils clean of mucus to facilitate breathing.

Once she has calved the cow will show a certain degree of excitement and desire to lick her calf. This assists the drying of the calf's coat and stimulates circulation. Rubbing down the calf with a wisp of straw or hay is recommended in addition.

Preventing Navel Infection

The navel of the calf should be powdered with a light dressing of copper sulphate (bluestone) or painted with tincture of iodine to promote healing and prevent navel infection, particularly if calving occurs indoors.

The herdsman should then see the cow has water to drink and should clean her teats before leaving her to mother her calf.

Heifers often take much longer to calve than cows and may at calving be more excitable. They should be disturbed as little as possible unless calving is abnormal.

A normal calving will occur within 20 to 40 minutes of the appearance of the water bag. If, after this time, the fore feet of the calf have not appeared, or the feet have not altered their position, an examination of the position of the calf is called for.

Two buckets of warm water should be at hand, with plenty of soap. Clean the hands and forearms thoroughly before this examination is made. Then carefully feel the position of the feet and ascertain if the head of the calf is in the right position.

When Trouble Arises

Three common complications here are that the calf may be coming backwards, in which case the tail can be felt but no head, or the head may be turned back, or one of the legs. In the latter cases the calf should be pushed back into the womb and the head straightened or the missing leg retrieved.

More difficult cases should be the immediate concern of the veterinary surgeon; ill-judged assistance given at this stage by pulling on the legs of the calf can cause great damage to the mother by setting up internal haemorrhage. If assistance is given by pulling on cords passed around each leg, make sure the presentation is normal before pressure is applied, and pull only when the cow heaves and always in a downward direction.

It is unwise to calve a cow tied by the neck with a chain.

If the cow has to be tied, a halter is preferable. Often a cow will lie down quickly from a standing position and may choke herself without a quick method of release—hence the advantage of calving in a loose box rather than in a cowshed.

Beef cows calving out-of-doors generally calve with fewer complications than dairy cows, but it is as well to be prepared and have them at calving time within easy reach of a crush in which they can be secured with the minimum of fuss while assistance is given.

Calving Complications

With cows the possibility of milk fever occurring should never be lost sight of. An unsteady gait before calving or paddling of the hind feet are warning symptoms, and the wise herdsman either calls in veterinary help at once or gives an injection of a calcium borogluconate solution himself. Cows may suffer from milk fever during calving and fail to calve promptly for that reason.

The afterbirth or placenta rarely follows the calf immediately, but it should be expelled within 12 to 24 hours. Its retention soon sets up a septicaemic condition in the genital organs.

The risk of this can be alleviated by inserting a pessary into the mouth of the womb immediately after calving and for every 24 hours until the afterbirth is expelled. But if it is still not expelled after 72 hours the veterinary surgeon should be called in.

On no account should unskilled hands pull on the membranes as the probability will be that they will break off and leave portions inside the uterus to set up inflammation of a most dangerous character.

It is my firm belief that much of the genital infection found in cows (such as metritis or leucorrhoea) is due to careless, unhygienic and ignorant attention to cows at calving time. The conscientious herdsman should leave nothing to chance and consult his veterinary surgeon whenever in doubt and before it is too late.

Make Sure Calf Suckles

Before the mother and her calf are left, the ability of the calf to stand and suckle should be confirmed. Some dairymen remove the calf once it has been licked dry by its dam and feed it with colostrum (the first-drawn milk) by hand, thereby causing less mental upset to the cow than if the calf is left to suckle.

Others leave the calf with the cow for up to four days—which is probably the wisest plan wherever calf mortality has been high as it ensures an adequate consumption of the all-important colostrum.

One final word on the question of calving and its relation to health. It should be part of the herdsman's routine duties to keep up-to-date and accurate breeding records. Only then is he able to confirm abnormal breeding behaviour and to know when calvings are to be expected.

This matter was discussed in greater detail earlier in this book, but its importance in creating the right approach to the control of disease and ill-health on the farm is too great for it to receive no mention at this stage.

Tackling Troubles on the Farm

Now we come to more detailed matters of tackling troubles and disorders on the farm. For ease of reference I have grouped them under the main recognisable symptoms, beginning with first-aid treatment for injuries.

Bleeding

To stop bleeding make a pad of lint or cotton wool (or part of a linen sheet or even a handkerchief) moistened in warm water and hold it firmly over the wound until a bandage can be applied to keep the pad in position and assist clotting of the blood.

Call for professional veterinary aid immediately, particularly if the blood tends to spurt out, indicating that an artery has been damaged. Keep the animal as quiet as possible.

Wounds

Clip the hair from around the wound and then bathe with a salt solution (1 teaspoonful of salt to 0.6 litre of boiled water) to assist in cleaning up the wound.

Damaged teats in particular should be promptly treated, and veterinary assistance called in immediately where wounds are deep-seated or badly contaminated with dirt.

A dusting with sulphonamide powder is a wise precaution where wounds cannot be bandaged; this keeps possible infection to a minimum.

Bruises

Bruises—including swollen joints—should be bathed with cold water on the first day and then with fairly warm water, at about 49° C, frequently on succeeding days.

Foot Injuries

The feet are the most usual seat of injury. The first job is thoroughly to clean the foot, paring away any surplus horn which may reveal a foreign body (a stone or nail) embedded in the foot tissues. On removal of this the site should be disinfected and the foot bandaged to keep it clean pending skilled veterinary treatment.

Choking

The best first-aid treatment is to try and locate the obstruction in the gullet by feeling along the whole length of the underside of the neck. If an obstruction can be found, massage towards the throat will in most cases dislodge the obstruction.

More stubborn cases require veterinary aid.

Damage to the Eyes

Cattle sometimes get pieces of chaff in their eyes, causing inflammation and a discharge very distressing to the victims.

The pain can be eased pending skilled veterinary treatment by catching the animal and holding its head on one side so that a few drops of warm castor oil can be dropped onto the eyeball from the point of a pencil or thin glass rod.

Hoven or Bloat

In this condition the left side of the cow becomes greatly extended between the last rib and the hook bone or hip joint; breathing is laboured, and prompt action is needed.

First, give a drench of 28 ml of turpentine in 0·6 litre of boiled linseed oil and cause the animal to take exercise. Massage of the extended rumen may help in the expulsion of gas by belching.

In cases which fail to respond to this treatment the use of a trocar and canula to puncture the rumen midway between the last rib and the hip bone wall is necessary. Failing this use a sharp knife, twisting the blade after insertion to allow the escape of gas.

Bloat occurs in a frothy form also where escape of gas is prevented by a frothy condition of the rumen contents. This condition is much more difficult to alleviate; the "spilling over" of the rumen contents into the body cavity between the rumen wall and the skin will lead to further complications unless veterinary aid is sought immediately.

To prevent bloat, feed a little dry roughage (hay or straw) before cattle graze short lush clovery pasture or lucerne. If strip-grazing, allow more frequent meals so that the cattle are not turned on very hungry with a tendency to gorge themselves. Under such conditions the use of a lie-back field of older grass is a definite safeguard.

Never graze cattle on very wet or dew-laden herbage. Mowing of lucerne in advance of the electric fence, allowing a 24-hour wilt before grazing off, is another worth-while preventive.

Coughing

May be due to several causes:

(*a*) *Infection with husk*

Usually noticed in young calves or yearlings towards the end of the grazing season. Coughing is aggravated when driven.

House such stock immediately and feed generously, or move to a new ley *not* previously grazed by cattle of any description and supplement with additional feed (see page 110). Oral vaccination is now available to prevent husk infection.

(*b*) *Infection with tuberculosis*

A dry, hard cough was symptomatic of TB infection of the lungs. Fortunately, with the elimination of tuberculosis from our cattle it is not expected to occur on farms today.

(*c*) *Bronchitis* (*respiratory chills*)

Usually accompanied by nasal discharge. Good nursing is imperative, house in a warm dry box, rug if necessary and give easily-digested palatable foods such as bran mashes.

Avoid drenching if at all possible.

Ringworm

Occurs most frequently on calves and yearlings housed under

unhygienic conditions or inadequately nourished. The source of infection is from fungal spores resident in wooden fixtures or dirt adhering to walls or tubular steel fittings.

Thorough disinfection of calf premises and treatment of woodwork first by a blow lamp and then by creosoting, should be made an annual routine.

For treatment, wash the affected parts with soft soap, scrubbing lightly to remove the scabs, and then dress with a mixture of 1 part creosote to 7 parts linseed oil, or in stubborn cases use old tractor sump oil. Aerosol sprays are also now available and are easier to use, particularly around the eyes.

Lice

Can cause unthriftiness, particularly in young cattle, and loss of hair. The trouble is generally associated with conditions similar to those accompanying ringworm infection.

Treatment is to dust the neck, backline, tail root and around thc top of the head and ears with louse powder obtainable from veterinary chemists.

Scouring or Diarrhoea

In its simplest form scour may be due to a stomach chill or change of diet, both capable of remedy.

Chronic or persistent scouring may be due to diseases such as Johne's disease or worm infestation.

The latter is amenable to treatment with anthelminthic, the former are not. If, therefore, no response is obtained to dosing with anthelminthic, immediate slaughter is advised and thorough disinfection of the premises. Otherwise infection is further disseminated.

Worms

Parasitic infection of young cattle is more common than has been generally realised. Routine drenching with anthelmintic is becoming standard practice on intensively-stocked farms.

Refusal of Food

Where no disease complications exist this may be due to overfeeding with unsuitable foods (e.g. frosted roots) and consequent indigestion, or to dirty feeding troughs or unpalatable foods.

If coaxing with a change of feed (after cleaning away all previous food residues) does not induce feeding then suspect other causes, for example—

By taking the animal's temperature (see page 116) the presence of a fevered condition may be established, associated with a chill, or inflammation of the udder or other organs. Veterinary attention is then essential.

Loss of Appetite

When the breath smells (as in acetonaemia—see page 127) the loss of appetite is probably associated with this cause. A dose of molasses 0·6 litre in 1·15 litre of warm water) is then advised.

To restore appetite in the case of digestive troubles 110 g of medicinal glucose in 0·6 litre of vinegar is helpful. A bran mash made crumbly moist with warm water, to which some molasses have been added and a pinch of salt, is also a useful appetising and pick-me-up feed.

A further cause of lack of appetite may be the presence of a foreign body (such as a nail or piece of wire) in the fourth stomach. Such a condition calls for expert diagnosis and treatment.

Failure to Thrive

Some cattle even on good feeding may fail to thrive, the chief causes being:

(*a*) *Parasitic worms.* Dose with anthelmintics as manufacturer's instructions.

(*b*) *Liver fluke infestation.* Dose with appropriate anthelminthic.

In this and in the above case there may be dropsical swellings under the jaws, anaemia of the eye membranes, and a harsh, staring coat free of lick marks.

(*c*) *Johne's disease.* Generally diarrhoea is present, but cattle can show no scouring symptoms and still be infected.

(*d*) *Lack of some essential mineral.* Shortage of copper or cobalt causes unthriftiness and can be combated by feeding, or allowing free access to, a complete mineral mixture in powder or brick form.

A depraved appetite (eating soil or de-barking trees) is often held to be a sign of mineral deficiency.

(*e*) *Excess of certain minerals.* Minerals such as flourine or molybdenum may be in excess in the herbage. The latter condition is the cause of "teart" pastures in Somerset and elsewhere. It is counteracted by feeding copper sulphate added to cattle cubes. The former is a result of air pollution from industrial processes and causes damage to the teeth and lameness.

Loss of Milk Yield

There are a number of possible causes of this. Among the most likely are:

(*a*) *Indigestion* resulting from unsuitable foods, or excessive consumption of certain foods, or a rapid change of diet, all affecting the activity of micro-organisms in the rumen or first stomach of the cow.

All dietary changes should be made gradually—over the space of four to seven days.

(*b*) *Acetonaemia*, which is a complex condition usually associated with loss of appetite, and with milk and breath both smelling of acetone.

Treatment is best left to a veterinary surgeon, but the incidence of this disorder can be reduced by maintaining the cow's intake of sugar during the first 12 weeks of lactation by the use of molasses given in a bran mash as a daily feed—see page 80. An alternative is to include molassed beet pulp in the diet.

Overfeeding with concentrates and inadequate consumption of hay or other roughage, or lack of exercise and green food, also appear to be contributory factors in this disease.

(*c*) *Mastitis* or inflammation of the udder may occur in mild (chronic) or in acute forms.

In the former, small flecks in the fore-milk may be seen when the first-drawn milk from each quarter is directed on to the blackened top of a strip cup.

In acute cases the fore-milk is much more pus-like and the quarter more inflamed and painful. These cases are much less amenable to penicillin treatment and often result in the loss of affected quarters.

Treatment of the milder forms of mastitis can now, thanks to the introduction of antibiotics, be undertaken by an intelligent herdsman under the direction of his veterinary surgeon, but he must realise the need for the strictest hygiene in infusing any teat with an antibiotic.

The teat orifice must be thoroughly cleaned with surgical spirit before the tube containing the antibiotic is inserted.

Acute cases of mastitis such as summer mastitis are best left to a veterinary surgeon.

(*d*) *Overstocking of the udder*

A long milking interval, resulting in the reabsorption of milk from the udder as a result of the overstocking, can if persisted in seriously reduce the milk yield; hence my recommendation for three-times milking of high-yielding cows (or milking at 12-hour intervals) where labour conditions permit.

(*e*) *Presence of disease*—many diseases such as anthrax, foot-and-mouth disease, bronchitis or pneumonia, lameness (foul-in-the-foot), husk or fog fever, and inflammation of the genital organs as a result of a retained afterbirth can all cause loss of milk.

In any case of doubt obtain veterinary advice without delay.

Infertility

A low rate of fertility, rather than actual sterility, is a problem in some herds or in individual cows within herds.

Where it is a herd problem, with the conception rate well below the normal 60-70 per cent of conceptions to first services, the most likely cause is nutritional—either gross underfeeding or a mineral imbalance. In the latter case analysis of the bulk feeds may reveal a mineral deficiency which can be corrected by supplementary mineral feeding—the most likely deficiencies today are in phosphorus, copper or iodine.

With individual cows, low fertility may be due to one or more of the following causes:

(a) Venereal disease, or inflammation of the genital organs as a result of retained afterbirths. This often happens with contagious abortion infection, or where infection has arisen as a result of lack of hygiene at assisted calvings (see p. 121). Cows with any genital discharge, except when on heat, should not be mated until examined by a veterinary surgeon.

(b) Mating at the wrong time. Too early service after calving is unwise—the chances of conception are much more favourable 8-9 weeks after calving than with earlier service.

(c) Quiescent ovaries. A cow may fail to come on heat owing to the ovaries being inactive—massage of the ovaries per rectum may overcome this. Generally this is best done by a veterinary surgeon but with practice a herdsman can master the technique.

(d) Insemination too early or too late in the heat period. The most favourable time is when the cow is going off heat—there is a better chance of conception in the 10 hours after heat than in the first 6 hours on heat. In winter the heat period may be very short, so turn out cowshed-housed cows every day to detect any cow on heat, otherwise she may be missed.

Accurate breeding records are of great value in assessing the fertility position in a herd, milk recording has the additional advantage of ensuring this. Finally, if in doubt as to whether a cow is in calf, your vet will help you with a pregnancy diagnosis after three months of pregnancy.

Sudden Death

Rapid death may be due to poisoning in which case symptoms of acute abdominal pain, scouring, or convulsions usually precede death.

Dangerous materials to which cattle should be refused access are: chemical spray cans, old paint drums, painted woodwork, particularly if lead paint has been used; and old drums which contained arsenical sheep dip. Pastures which have been sprayed with hormone weedkillers should be left unstocked for 10 days.

Some weeds are poisonous to stock either eaten green or in hay, notably ragwort, hemlock and yew.

Mild poisoning may be induced by excessive consumption of certain foods such as fodder-beet, sugarbeet tops, rape, and green potatoes. Beet tops should be fed after wilting, and (as with fodder-beet and rape) introduced gradually into the diet, with some dry food (e.g. hay) available to the stock. Green potatoes should be steamed or boiled before feeding.

Pastures which have been top-dressed with fertilisers, including basic slag, should not be grazed until the herbage has been washed clean by rain.

In all cases of sudden death the advice of a veterinary surgeon should be sought at once. Anthrax (a notifiable disease) is often responsible for sudden deaths, and to interfere with the carcase in any way is highly dangerous as the blood will be heavily charged with anthrax germs.

Another cause of sudden death is 'Black Leg,' particularly in younger cattle. After death the tissues swell up to assume a puffy texture and the skin becomes parchment-like. Where the disease has been known to occur, all young cattle should be inoculated with vaccine by a veterinary surgeon every spring before turning to grass.

The disposal of carcases from such cases as the above should always be the subject of veterinary advice to reduce the risk of spreading infection.

Hypomagnesaemia

Another cause of sudden death is hypomagnesaemia, or lactation tetany, characterised by magnesium deficiency in the blood.

Pre-disposing factors include adverse weather or exposure. Generally occurs with cows in-milk or suckling, particularly on diets low in magnesium—such as young spring grass,

Better than Drenching

SOME veterinary surgeons advise mixing powder medicines with jam and smearing them inside animal's mouth with smooth piece of wood. Fluid medicines can be mixed with calf gruel powder and administered similarly.

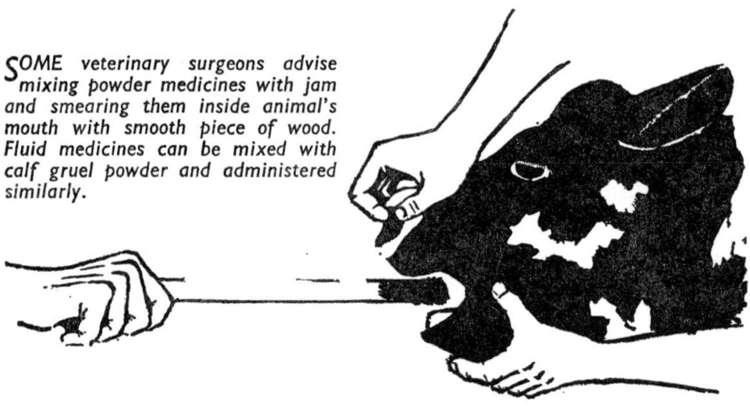

especially if top-dressed with sulphate of ammonia, or on soils abnormally high in potash.

The symptoms exhibited before death, which may be very sudden, are excitability, muscular tremors and an uneven gait. Prompt treatment by a veterinary surgeon, involving an intravenous injection of a magnesium solution, can effect a dramatic cure. But cases occurring in the night often terminate fatally.

Preventive measures (see page 40) involve the feeding of magnesium-rich mineral mixtures, or including 60g of magnesium oxide in a carrier food such as crushed cereals or dried beet pulp fed to cattle at pasture. Hay feeding during the early part of the spring grazing period (when hypomagnesaemia is most common) also helps.

Where silage is fed, the addition of 1·8 kg of magnesium oxide (calcined magnesite) to each tonne of grass ensiled is advisable.

Potassic fertilisers should be applied to pastures in the summer or autumn, rather than in the spring. My own practice is to use potassic superphosphate on all pastures as an autumn top-dressing and apply nitrogen only as the spring fertiliser, breeding ewes being affected as well as cows.

Disease and the Law

To conclude this chapter mention must be made of diseases which are in some measure controlled by legislation. They are:

Warble Fly

Legislation is currently proposed that all cattle showing signs of infestation with the walnut-size swellings along the back (typical of warble fly attack) should be treated in the spring.

Warble fly attacks in summer cause much loss by cattle "gadding" and by reducing the value of hides for leather manufacture.

Of recent years two new lines of attack have been developed against the warble fly, involving destruction of the larvae during their migratory period in the body of the host.

Proprietary insecticides are now strongly recommended for brushing on the backs of store cattle in the early winter, November, and in the spring. One course of treatment only is required and, as compared with derris dressings, it saves time. Gathering cattle in April/May for treatment is often inconvenient at a busy time of the year.

Notifiable Diseases

Should any of the following diseases be confirmed on the farm the disease must be notified immediately to the police, or to a veterinary surgeon:

Anthrax. Symptoms—sudden death; or, if less acute form occurs, high fever—temperature 41-42° C and prostration.

Foot-and-mouth disease. Symptoms—loss of appetite, lameness; later dribbling from the mouth, blisters form on the tongue, gums, teats and feet.

Tuberculosis of the udder. Symptoms—hardness of the lymph glands high up in the rear of the udder. (N.B. Mastitis may cause hardening of the udder in a more generalised form.)

Sale of Milk

It is illegal to sell milk from cows undergoing medical treatment. No milk from cows treated with antibiotics must be sold until at least 48 hours have elapsed (i.e. four milkings must be withheld).

CHAPTER IX

REARING AND MANAGING BULLS

BULL calves are usually reared on a high plane of nutrition. This usually allows the full growth potential to be realised, it permits of earlier use for breeding and ensures that, for licensing purposes at 9-10 months old, the young bulls are presented in good condition.

Methods Used

Of the methods employed, single suckling is most popular; but it is, naturally, expensive.

In some of our pure beef breeds, where the milk yield of the dam is often very low, a foster mother of greater milking capacity is often employed. Such a practice, in the long-term view, is to be condemned as it may well lead to strains of beef cattle being bred incapable of rearing their own calves under range conditions of single suckling.

However, multiple suckling is commonly used for bulls of our dairy and dual-purpose breeds where the development of early-maturing, fleshing qualities is of less importance than in pure beef breeds.

Pail feeding is less common in rearing bulls, but in skilled hands can be quite satisfactory.

Details of calf rearing have already been given in chapter VI. The tables given there—on pages 91 and 92—apply to steer and heifer calves, but are quite satisfactory as a guide for bull rearing provided that the level of concentrate feeding is stepped up to 25 per cent above the levels shown, and provided that weaning takes place later for bulls than for steers or heifers.

Handle Him Often

A young bull cannot be handled too frequently. He should be taught to lead on a halter from an early age. Be patient and firm in handling him and speak to him to gain his confidence. Never play with him about the head.

By the age of nine months a bull should be rung. Tie his head up firmly, then punch a hole in the septum of his

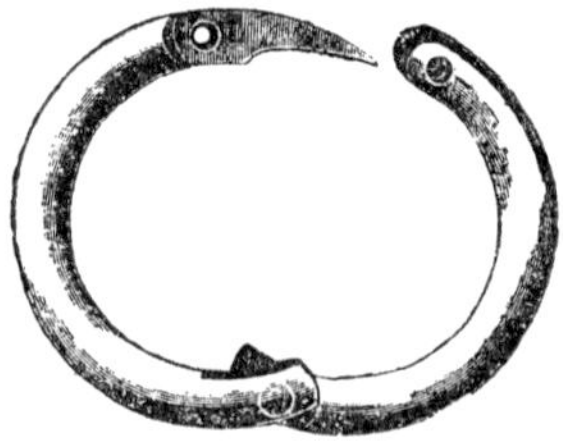

Above: Bull nose punch and, left, ordinary nose ring. Right: Self-piercing ring.

nose (using a bull nose-punch) to admit a copper ring of 50-60 mm diameter. Self-piercing bull rings can be bought.

Once the nose has healed the bull should continue to be led with a halter; another short piece of rope attached to his ring will enable his head to be held high.

With older bulls a bull staff is often used. This enables much better control to be exercised, as a bull charges with his head down.

As sexual maturity is reached by 9-10 months young bulls are generally boxed separately from that age, but should still be exercised and groomed daily.

Housing for Bulls

With advancing age a bull requires regular careful handling, and a properly-designed bull box and exercise yard are essential if bulls are to be kept to the age of five years or more which is necessary if progeny testing, so fundamental to long-term breed improvement, is to be carried out.

The essentials of good bull accommodation are:

1. A strongly-constructed loose box with smooth inside walls with a built-in feeding trough in which feed can be placed from a feeding passage or through a feeding hatch without the herdsman having to enter the box.

2. An outside exercise yard with tubular steel rail walls into which a service crate is built. The cow can be let into the crate through a gate which shuts off the crate from the bull yard.

Ensure Easy-Catching

Various methods are used to enable a bull to be caught easily when he is in his pen.

One method is to keep the bull attached to an overhead cable by a chain passing round his horns and through his ring. This tethering chain should be fitted with several swivel links and be equal in length to the height of the overhead cable above ground level. If it is too long it may foul the forelegs when the bull lies down, if too short the bull cannot lie comfortably.

The overhead cable should run from above the feeding trough in the box to above the service crate so that the bull can serve a cow without being released from the chain. This method ensures greater safety to the herdsman in that the bull can be caught and securely fastened when his box has to be cleaned out and littered.

Another method is to use a yoke manger in which the bull can be secured once he puts his head through the yoke to feed.

But unless a bull is trained to this method when young and becomes accustomed to it he may become too wily to be caught —even though the yoke mechanism is operated from outside the box.

A third method is to attach a short length of chain—short enough not to interfere with feeding—to the bull's ring, and rely on catching him as he feeds.

To avoid doing this every time he is cleaned out, it is a most useful device to have a vertically sliding door or tubular steel rails which can be inserted to cut off the box from the exercise yard and so be able to clean out either section without risk.

Well-fed, Not Over-fed

Bulls, to be fertile, must be adequately fed but should not be over-fat.

Many young bulls, particularly in the beef breeds, are sold in such high condition that they need to be reduced to a leaner and harder condition before service; whereas older bulls are often under-nourished, and low fertility due to low virility in the semen may be the result.

Feeding Guide

During the height of the breeding season a bull should be better fed than during the off season if conception rate is to be satisfactory. The following is a feeding guide for mature bulls:

	Bulls in active use	Bulls only sparingly used
Winter		
Hay	1·3 kg per 100 kg liveweight	0·9 kg per 100 kg liveweight
Straw—	None	0·9 kg per 100 kg liveweight
Silage, kale, roots or wet beet pulp—	1·8 kg per 100 kg liveweight	1·8 kg per 100 kg liveweight
Concentrates—	1·5-3·5 kg	None
Summer		
Hay—	0·45 kg per 100 kg liveweight	0·45 kg per 100 kg liveweight
Grazing on tether—	4 hours	4 hours
or green forage—	25 kg	25 kg
Concentrates—	1·4 kg	None

Suitable concentrate mixtures are:

Food	*Winter* per cent	*Summer* per cent
Cereals (oats, barley, maize)	50	75
Linseed cake	25	25
Bran	20	
White fishmeal	5	

(Approximates to a balanced dairy cake)

Younger bulls still making active growth are generally fed along similar lines to the above but will receive 1·5-2·0 kg concentrates every day until maturity, which will be reached by approximately 2½-3 years of age.

The First Service

At his first service a young bull may be easily frustrated. He should be turned into a paddock or a yard with two or three heifers for service and allowed to "play" with them to arouse his sexual desires. (Some young bulls heavily fed and suckled on nurse cows may show perverted desires and be very slow to serve.)

The heifers should be chosen to be of a size which makes service easy. Then once the bull has served successfully he can be used on larger and older females.

In cases where there is a big size disparity the females may be held in a crate and their hindquarters lowered by digging out a pit for them to stand in.

The most effective stimulus for a young bull is for him to have a view along the cow's back and this is provided by standing him on higher ground.

Why Bulls May Become Slow

Older bulls may be slow in service owing to inhibition by fear of pain or injury or strange surroundings. Some old bulls, for instance, refuse to serve in a crate if unaccustomed to it, and a slippery floor may make mounting difficult.

Overgrown feet, or a sore brisket may be reasons for slowness in service.

Very occasionally a bull may injure the penis in serving and may need veterinary attention. It is always a wise plan to remove any dried dung from the tail of a cow or heifer to be served as such accumulations can cause lacerations of the penis with an eager bull.

As bulls grow older they often become increasingly slow, and often then the best plan is to turn the bull loose with the cow in a well-fenced paddock or exercise yard.

During the "off season" a bull can be kept much more even-tempered if he is run with some dry in-calf cows; deterioration in the temper of old bulls often results from loneliness and frustration if confined to a box.

After a period of inactivity a bull may temporarily produce low virility semen and should therefore be permitted more than one service.

Advantages of Tethering

To tether a bull during the summer is a commendable practice. It affords exercise, and the grazing is beneficial as well as overcoming boredom. In very hot weather, when flies are troublesome, avoid tethering during the heat of the day.

Special bull-tethering equipment is now available, or the bull can be secured by a chain round his horns through his ring to a stout cable anchored at each end. The tethering chain can travel the cable if secured to it by a large link.

Polled bulls can be tethered by fixing the chain to a stout leather head collar.

Bad-tempered bulls may need restraining by the use of a mask, particularly when being led; a bull pole or staff should always be used in such cases for the safety of the herdsman.

Foot Trimming

A bull's feet should be carefully watched as the hind feet particularly may need trimming. Plenty of exercise on hard surfaces helps to keep the feet in trim, whereas living on accumulated dung often leads to overgrown feet which, by throwing undue weight on the heels, may make service difficult and painful.

Feet can be conveniently trimmed in cattle stocks or after casting, using blacksmith's pincers or feet-trimming shears.

With cattle stocks, the bull (cows can be handled this way, too) is secured in the stocks, and a canvas sheet or surcingle is passed underneath the belly and tightened by rotation of rollers—one on each side to which the sheet is secured.

This effectively prevents the animal from lying down.

Each foot can then be pulled up in turn and fastened to a sill on each side of the crate below the roller. Trimming is then carried out, the sole of the foot being pared level with the cut-back wall of the foot.

It is a very sound plan to have the base of the stocks in the form of a shallow bath which can be filled with a 10 per cent solution of copper sulphate (bluestone) in which the bull can stand and receive foot bath as well as foot-trimming treatment.

Bulls are better handled in a crate than by casting; but the method of casting is to loop a rope round the neck just forward of the shoulders and carry the loose end down the spine, making a half-hitch round the body behind the shoulders and again across the loin or "small" of the back and just forward of the flanks.

One man holds the animal by the head and two more pull on the rope until it falls onto its side—the man at the head then holds the head down, one man keeps the rope taut and the other can trim the feet.

Disorders and Diseases

Bulls in winter often become infested with lice. I find it is advisable to clip the top of the neck, the withers and along the back and apply a louse powder dressing every three months, as the irritation set up by lice can cause bulls to be restless and destructive in their efforts to rid themselves of the irritation by rubbing door posts and fences.

Two serious diseases of dairy cows are spread by the mating act—vibrio foetus and trichomoniasis. Once a bull is infected with either of these after serving an infected cow, he should be regarded as a source of infection should he be mated to clean females.

Both diseases are characterised by abortion—often very early in pregnancy when the foetus is so small as to pass unobserved. All that may be observed is a trace of slime on the tail and the subsequent return on heat of a cow thought to be three or four months in-calf.

In such cases do not mate the cow with any bull until veterinary advice has been obtained.

No young bull should ever be mated to a cow that has been known to abort or has had an unsatisfactory breeding history. He should be used only on heifers at first. As he grows older, by confining service to females of his own age or younger, the control of these venereal diseases will be greatly facilitated.

Bull Licensing Arrangements, except for AI, have ceased.

Breeding Policies

Many herdsmen today are concerned not only with day-to-day problems of bull management but work in some degree with their employers in the larger question of herd breeding.

Because of this I think it will be helpful to consider now some of the many aspects of that work.

It is often questioned whether bull licensing is a means of breed improvement. The answer to this question is that bull licensing in its original form was a method of breed improvement—but in a negative way only. It eliminated the obvious scrub bull but did nothing to further scientific breeding. Its provisions were based on a false premise:

That judgment of a bull's conformation at ten months old is an accurate assessment of what the bull will transmit in conformation to his progeny. But conformation in the young animal can be largely determined by the level of nutrition—hence the practice today of rearing all young bulls on a high level of feeding to mask conformational defects.

Such a practice with dairy bulls may be entirely unsound.

With beef bulls, rearing on a high plane of feeding does enable the factors of early maturity and fleshing qualities to be revealed, and the criticism of judging on conformation alone is less valid. (See also page 143.)

Where conformational defects such as an undescended testicle, or an undershot jaw—both inherited defects—are revealed, then inspection serves a most useful purpose.

With the expansion of artificial insemination the demand for bulls is likely to fall unless their breeding is of such a high standard as to make them eventual candidates for use in the AI service or for them to qualify as register-of-merit sires in the respective breed societies.

THE AI SERVICE

Under the Agricultural (Artificial Insemination) Act of 1946 the Milk Marketing Board in England and Wales and the Scottish Boards have undertaken to provide AI facilities for farmers, including those who are not registered milk producers.

Each centre has an Advisory Committee largely representative of the membership with a nominee of the Veterinary Medical Association in the area. A manager is in charge of each centre with a team of trained inseminators who visit the co-operating farms of the members of the centre when required.

Notification for a service is, where possible, given to the centre concerned before 10 a.m. when insemination is usually carried out that day. If notified after 10 a.m., insemination is usually done early the next morning.

Since 1945 the development of AI has been extremely rapid

—from 2,599 cows inseminated in 1944/45, to 1·84m first inseminations in 1976-77.

The MMB operate 16 main centres, and 6 private centres operate under licence from the Ministry of Agriculture. Farmers can operate their own AI under licence in their own or other herds, as is being done by a group of British Friesian breeders (Cattle Breeding Services Ltd).

Advantages

The cost of proving a dairy bull is very high but the AI Service makes exceptional sires available to all milk producers regardless of herd size.

The advantages of artificial insemnation are greatest in small herds where the services of an expensive bull cannot be fully utilised. Equally, on small heavily-stocked farms the use of AI frees accommodation and food for other stock.

The method of testing and selecting dairy bulls for AI has great potential for raising the production of the national dairy herd, especially when it is realised that the bulls used by the MMB of England and Wales may inseminate many thousands of cows.

INSEMINATION FEES FOR THE MMB
(England and Wales)

Type of Service	*Fee** £
Commercial (Bull of the Day)	2.45
Nominations:	
Dairy Progeny Test Bull	4.00
Standard	4.00
Premium	5.00
	7.50
	10.00
Red Star Beef	2.95

*This fee is for a first service and a free repeat.

All chargeable repeats (3rd, 5th etc) carry a service fee of £2.20 which, again, includes a free repeat service (4th, 6th etc).

Selecting Dairy Bulls for use in Artificial Insemination.

A bull's merit is assessed today by using a highly complex and sophisticated technique. Firstly, fertility is examined.

The semen produced must be visually acceptable and capable of being diluted extensively and recover well after being frozen and give a satisfactory level of conception as judged by the 30-60 day non-return rate of the cows' insemination.

Secondly, the bull's promise in respect of yield of milk and milk solids, per cent milk solids and its progeny's conformation are assessed by progeny test. The Ministry of Agriculture must appove the bull for up to 750 inseminations for this test and after these have been completed, it is laid off awaiting the results of progeny test of over 500.

Seriously abnormal calves reveal that the sire is a carrier of undesirable genes and eliminates it.

Daughters' yields are examined for quantity and quality. Quality is assessed by an indirect method based on fat and protein information. This system is known as the Improved Contemporary Comparison (ICC) which has the merit of eliminating the effects of management while allowing bulls to be compared with a constant base.

Thirdly, a panel of breeders, appointed by the elected AI Centre committee, together with an MMB livestock officer, assess the conformation of the progeny and also all the other heifers in the herd. Each character is assessed against a score of a 100, which represents the National Average for that character. Greatest emphasis is placed on udder shape, teat shape and placement and legs and feet.

Register-of-Merit Bulls

Most breed societies accept bulls as proven (register-of-merit) bulls if at least 10 daughters have reached the necessary breed standards in their first lactation.

Not all register-of-merit bulls are equally good; what is most important is the proportion of daughters reaching the required production standard out of his total progeny. This information is often hard to find, as unsatisfactory progeny may have been sold into herds which are not milk recorded.

There is no doubt whatever that progeny testing is fundamental to progress in cattle breeding.

To summarise in this most important matter of selecting the right bull calf to rear, in the light of existing scientific knowledge the best guidance is:

1. Choose the son of a sire with an already impressive progeny record. This means that the half-sisters of the bull calf should show high production, desirable conformation, milking temperament (quick milkers), and evidence of good wearing qualities (particularly in udder attachment).

Such sires are identified by reference to their progeny records in the MMB or breed society publications or in progeny testing station results.

2. Choose a calf from a cow from a *family* of high and consistent producers, preferably an old cow that has two or more daughters already in milk and of satisfactory type and production standards.

3. Make sure the calf is a worthy representative of its breed and then rear it well.

4. Finally, use a young bull sparingly until his own progeny record is known; on that his eventual wider use depends.

PERFORMANCE AND PROGENY TESTING OF BEEF BULLS

Together with visual assessment of the progeny of beef bulls, many breed societies are concerned with obtaining factual data on the breeding record of their sires. Scientific techniques based on growth measurements of both the bull himself and his progeny, together with carcase grading are used by private herds, breed societies and AI organisations to select the top bulls for breeding.

Performance testing involves weighing beef bulls during their growing period up to say 18 months of age, to determine their own inherited capacity for growth and fleshing ability. Such characters are inherited so that, in general, bulls which show high rates of liveweight gain (over 0·9 kg per day) are likely to produce progeny of similar growth capacity.

The degree of heritability varies, however, with different bulls. Therefore, to assess the transmitting ability of a bull for these characters, it is necessary to test his progeny, particularly if he is to be widely used in AI.

The Animal Breeding Research Organisation, the Meat and Livestock Commission and the MMB are concerned with recording growth rates of bulls and their progeny by periodic weighings while rearing under controlled conditions of feeding and management.

The MMB's beef progeny testing station at Lambourn, Berkshire tests 22 bulls a year of the most popular breeds particularly the Hereford, Charolais and Aberdeen Angus. Carcase classification is carried out using the MLC's system. The top bulls of each breed enter the Red Star Beef Service. The calves from these sires often fetch a premium in the market as they are likely to grow faster than their contemporaries. The MMB also conducts a postal survey of calvings following service by the bulls on test. In this survey, farmers are asked to fill in a card recording the calving, whether it was normal or difficult and if the calf itself had any abnormalities. This sort of information helps the MMB identify the top beef bulls and provide full information to the customers.

Do-it-yourself AI.

Many farmers are turning to DIY AI to supplement the AI technician service. They feel that this gives them greater flexibility of choice of bulls and time of service. Deep frozen semen straws are stored in the farmer's own flask and inseminated by a trained member of the farm staff. A number of organisations, including the AI organisations, provide a complete range of services for the DIY breeder from training programmes, equipment and flask servicing to actual semen sales.

Removing unwanted teats

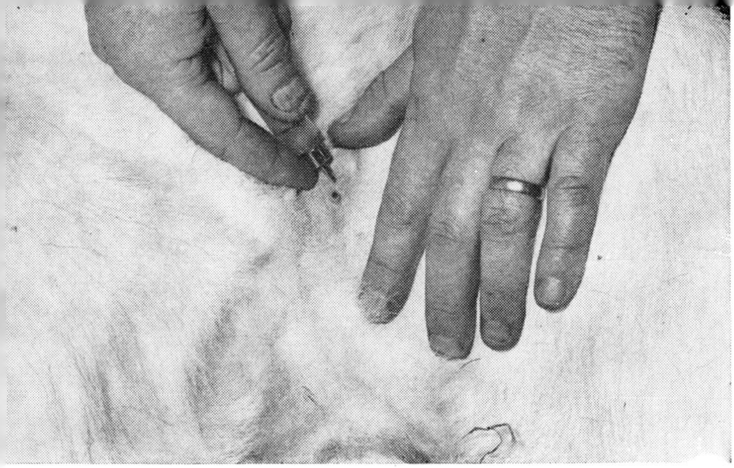

First step is to inject 0·5 ml of local anaesthetic under skin at base of each extra teat, after clipping hair short and washing area with antiseptic.

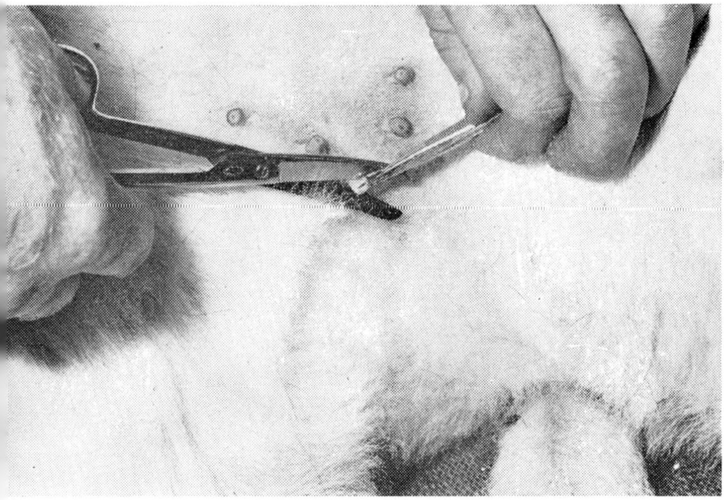

After waiting for at least one minute for anaesthetic to take effect, unwanted teat is pulled outwards as far as possible with pair of tweezers and cut off with curved-blade scissors, cutting well into teat base to remove rudimentary milk sinus.

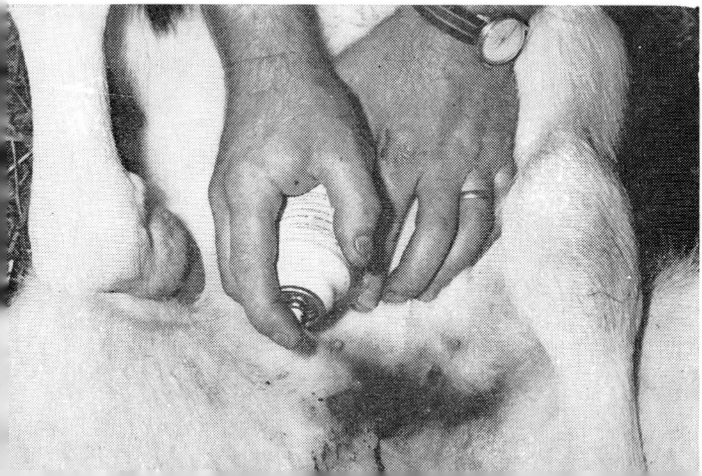

Finally, each open wound is dressed with antibiotic in cream, powder or aerosol form to prevent infection.

For the Burdizzo, method, testicles are drawn down to the end of the purse and one cord manoeuvred to the extreme edge—where finger is pointing.

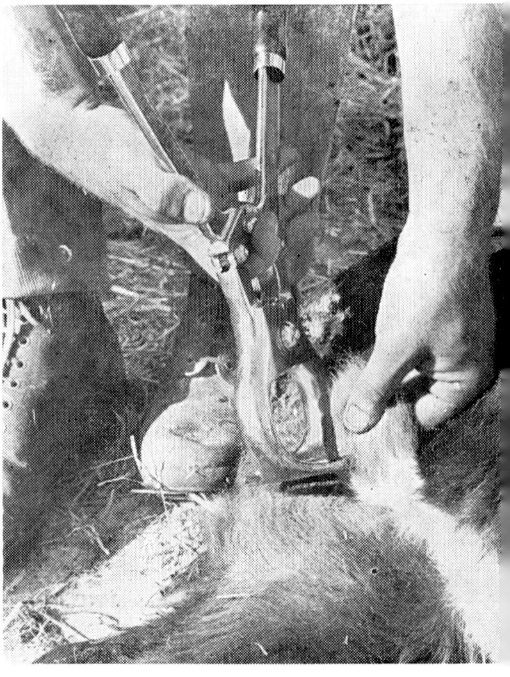

The Burdizzo is then clamped over the isolated cord and the handle compressed to the limit for 5 seconds. Object is to compress sides of artery and cut off blood supply to each testicle which then withers away.

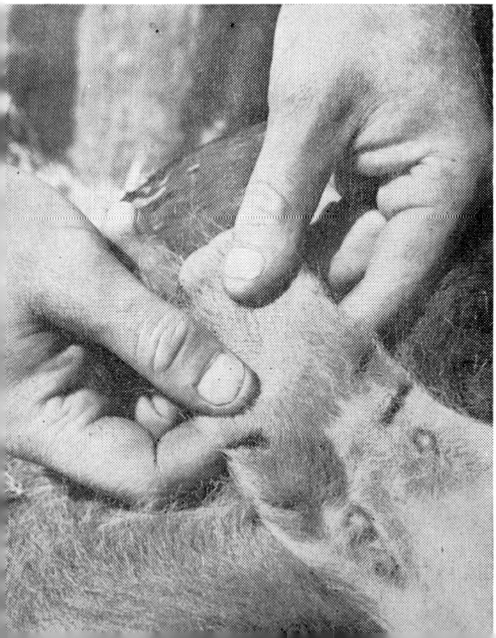

This is what the purse looks like afterwards—a double clench on each cord has been carried out to make sure. It is illegal to castrate a calf over 3 months old without an anaesthetic.

Castration methods

Alternative method for calves under 7 days old is a rubber ring applied with elastrator pliers. It is illegal to use this method after one week old.

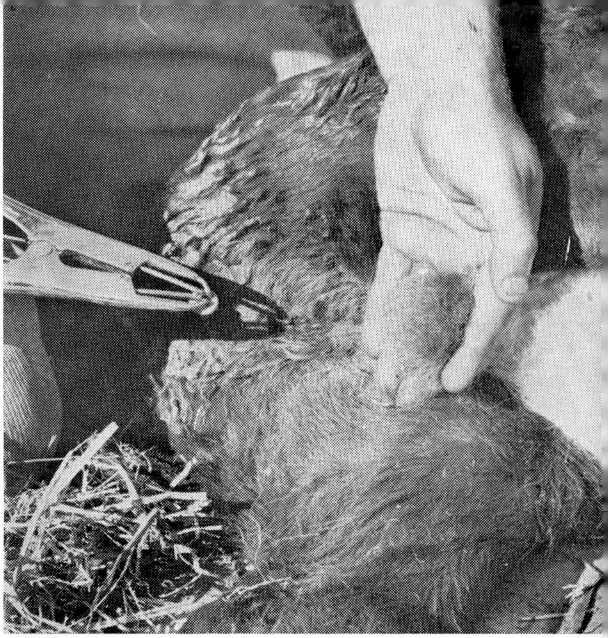

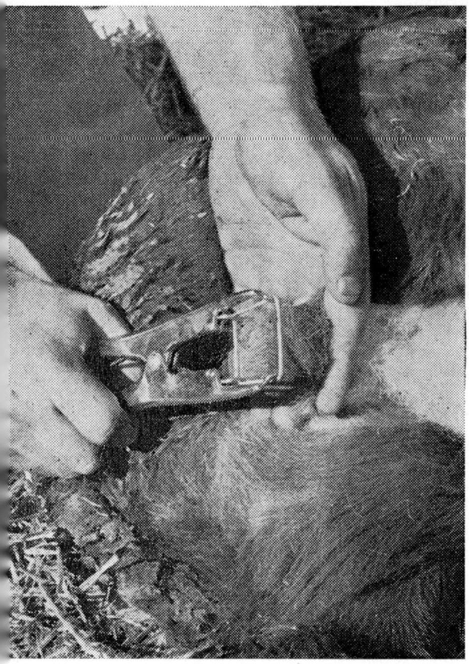

After drawing testicles down to end of purse, ring is stretched by opening plier jaws, slipped over testicles and worked up the purse.

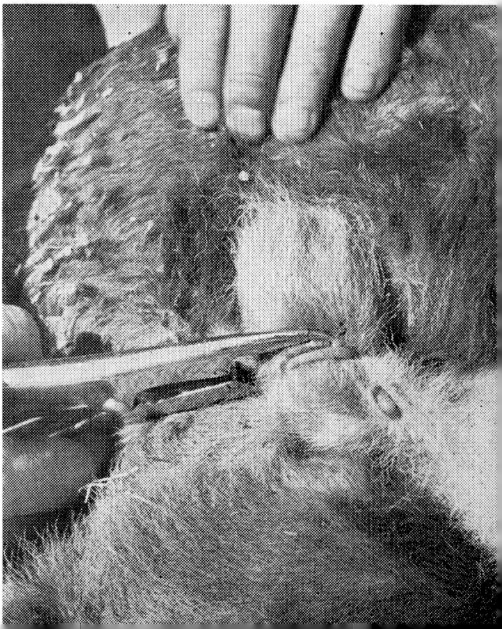

Job is completed by gently closing pliers and removing them from ring, after making sure that both testicles are clear below ring placed not too high up scrotum to trap testes.

Disbudding calves

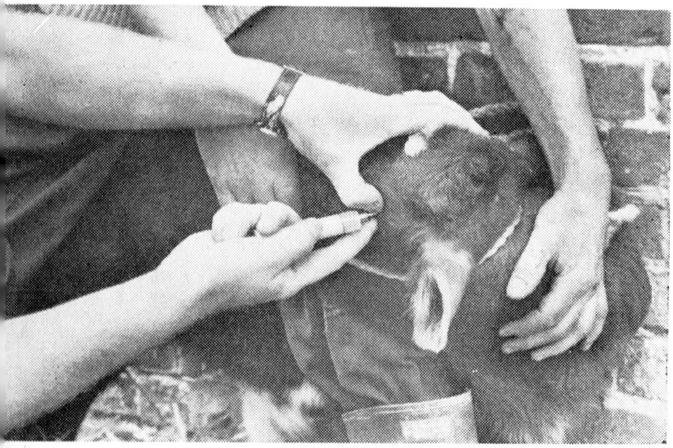

For disbudding with an electric dehorning iron, a local anaesthetic must be used to 'freeze' the nerves running to the horn buds. Training is necessary before this job can be undertaken.

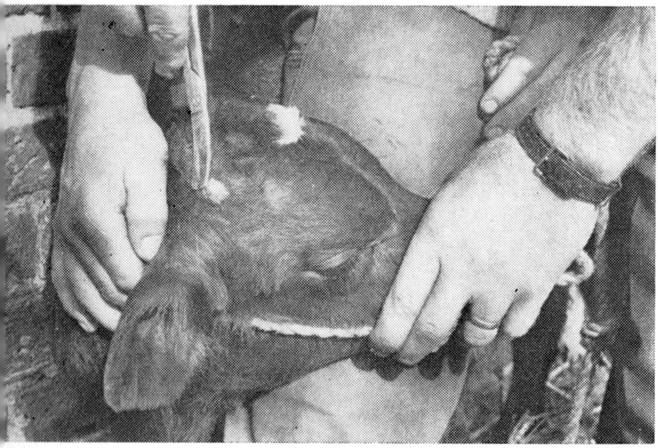

Hair round the horns can be clipped back to expose the immature buds during the 5 to 10 minutes necessary for the anaesthetic to take effect.

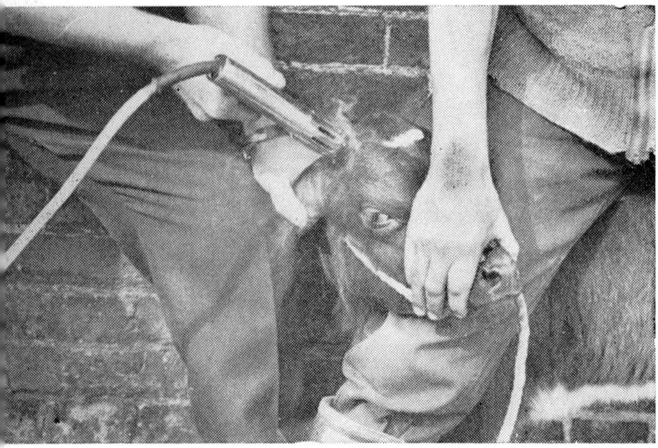

Heated dehorning iron is applied so that bud is fully enclosed in burning ring, pressed down, and rotated to speed burning action and remove any build-up of burnt material. Bud is removed by knife or flick of iron when burning reaches a layer of white cartilage.

Tattooing

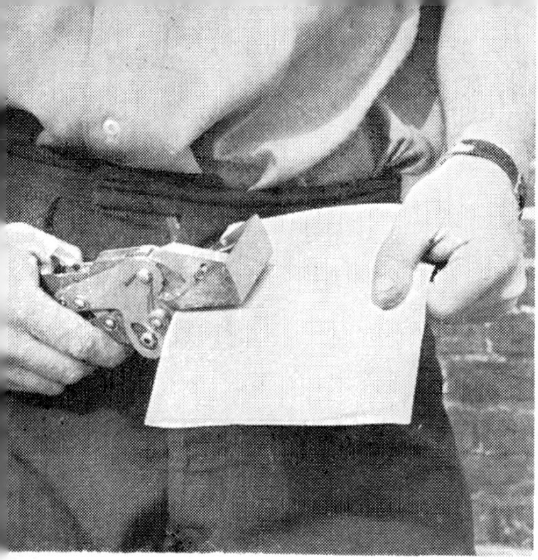

First job is to check that numbers and letters have been placed correctly in forceps by making test impression on piece of paper.

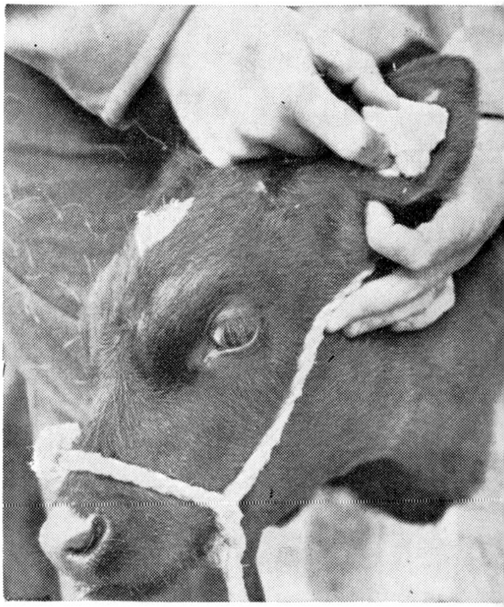

Calf's ear should be cleaned inside and out with cotton-wool swab soaked in spirit to enable veins to be seen and prevent dirt being introduced.

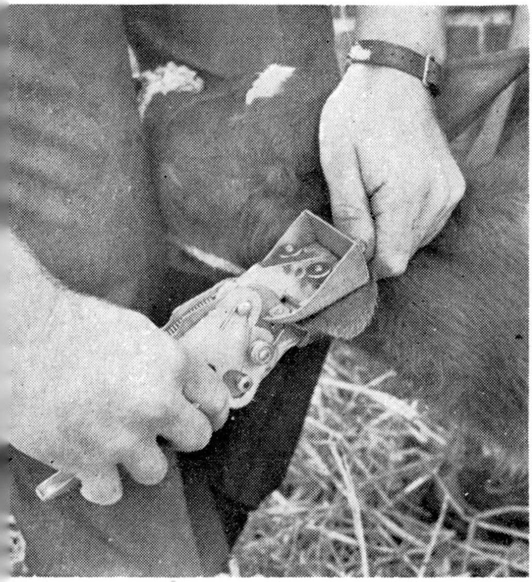

After applying tattooing paste sparingly to prongs with a brush, grasp leading edge of ear with one hand and apply forceps to inside of ear as near to tip of ear as possible to make for easy reading of number.

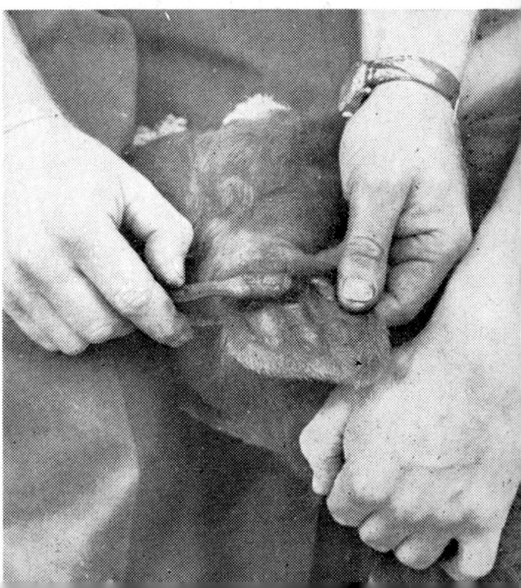

Finally rub tattooing paste well into impression with brush so that paste penetrates thoroughly to leave an indelible tattoo.

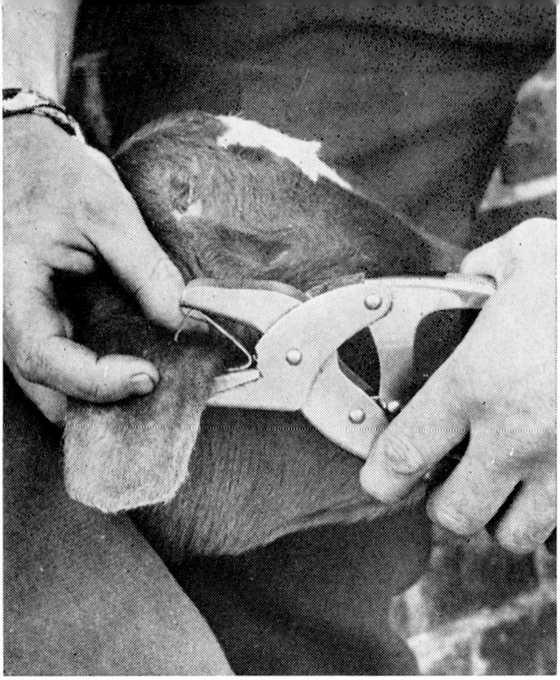

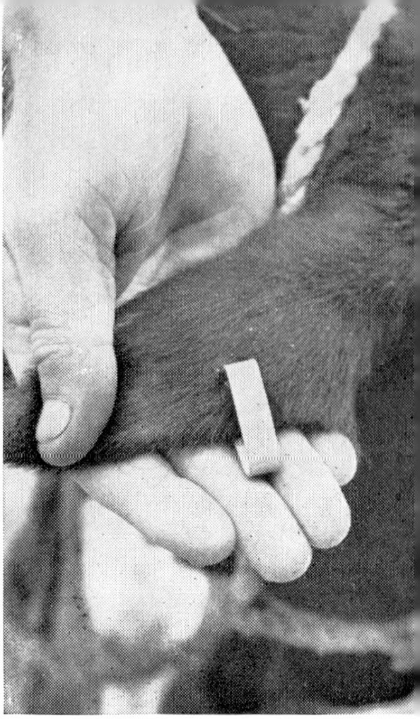

Metal tags should be clipped into the leading edge of the right ear, as close to the head as possible to lessen risk of being torn out but overlapping the leading edge by 14 mm to allow for growth, taking care to avoid veins.

Applying eartags

Two-part plastic tags are applied in the same position. With these there is no loop to get caught up—the parts just swivel round.

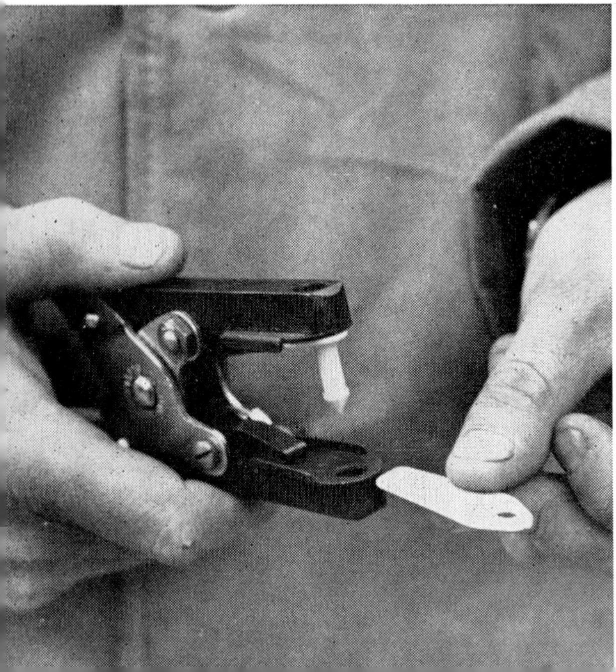

Clipping a cow

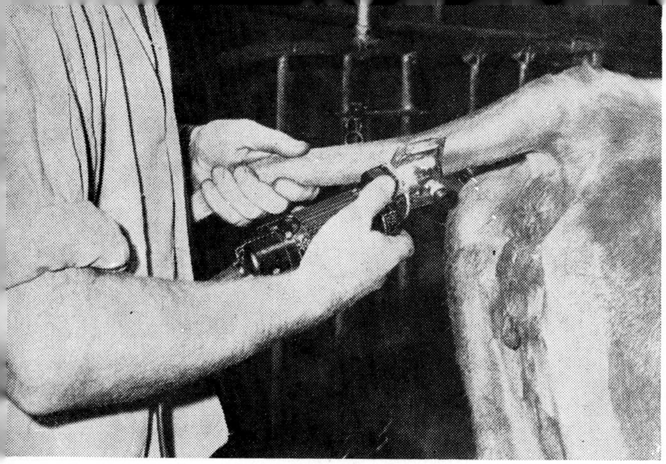

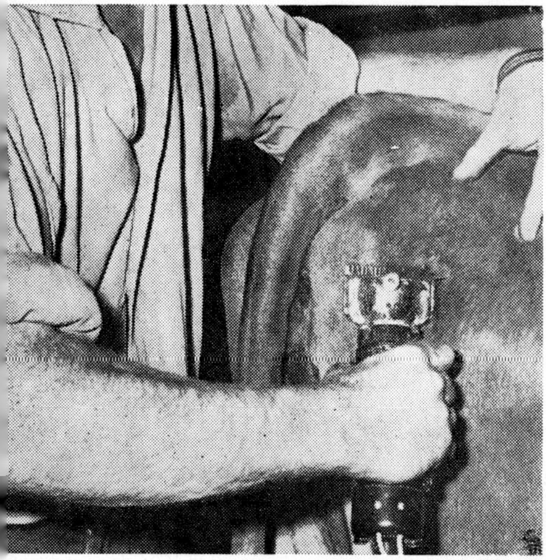

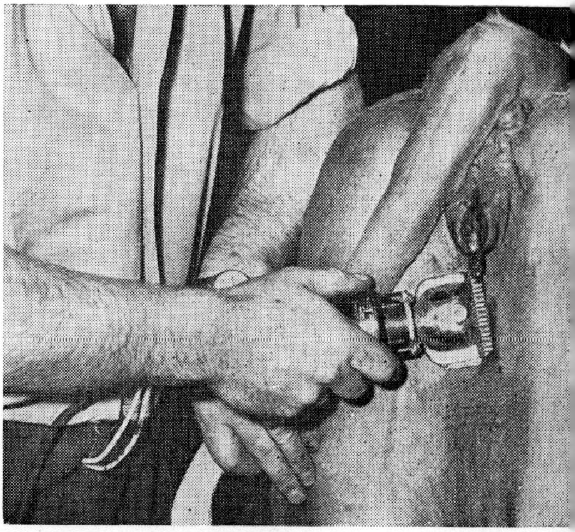

Begin by clipping tail close to the skin all the way up to the roots, leaving the switch.

Then clip a strip up to 300 mm wide on each side from the hock, up the leg and the back of the rump to the tail head.

Don't overlook clipping the escutcheon, where long hair can get soiled with excreta.

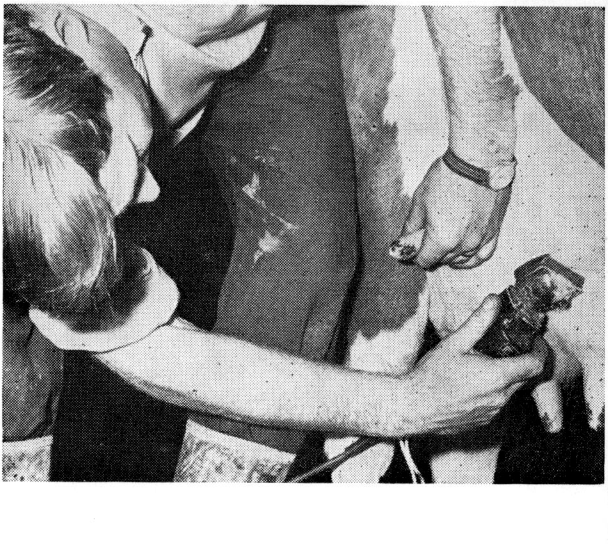

Finally, the udder should be clipped all over but not too close or it may get sore when washed before milking.

CHAPTER X

HANDLING THE HERD

As dairy herds increase in size, it becomes increasingly important to observe certain rules in the actual handling of them. Large herd management demands great skill and care on the part of the herdsman to ensure that the advantages of scale and mechanisation are not sacrificed through the achievement of a lower level of performance of the individual cows.

Accident Risk

Unless adequate precautions are taken and care is exercised, accidents per 100 cows occur more frequently in large herds than in small ones.

Every accident, no matter how trivial, should be recorded and investigated because behind every accident there is a cause and, as far as is possible, the cause should be identified and recorded. Often accidents of very wide variation arise from a common fundamental cause. It is far better to remove the causes of troubles than to rely on curative treatments.

Investigations will often point to the need for changing a routine, regrouping the cows, altering water trough sites or road-surfacing materials, or selecting different fencing materials or fence-line layouts.

Accident-prone herdsmen are expensive and, as a last resort, complete retraining may be the only salvation. This involves a complete break-down of the work in fine detail and retraining in every aspect—a formidable task. All accidents, however, arise not from the operative so much as from circumstances imposed on the herd, which are themselves perennial risk-elements.

Complete co-operation of the operatives is essential in such investigations and the management must make it clear that all parties—the operatives, the cows, the circumstances and the management—are under objective scrutiny. The exercise is not to apportion blame but rather to identify and remove the cause.

Dangers of Rigid Routines

When herd size is greatly increased, the routining of all operations tends to be carefully scheduled and the operatives can become routine-dependent and unable to deal with exceptional cases in exceptional ways when they first recognise a small abnormality.

There are two dangers in this situation. First, the decision may be taken to delay special treatment in the interest of carrying through the routine. Second, the faculty for identification of abnormality may be lost, especially if standards are allowed to sag even slightly in the interest of expediency in executing a rigid routine.

In the early days of operating a new installation or new accommodation, everyone involved is alert and interested to spot what modifications are necessary and these are attended to promptly and the animals are extra-carefully tended.

As time passes, small pieces of equipment may fail to function properly, corners are cut and perhaps short-cut methods are introduced and troubles build up hardly noticed and yet insidiously. Blocked drains, damaged ventilators, smooth concrete surfaces and a hundred other items are responsible for major losses in dairy herd function unless vigilant care is taken to observe them and to get them attended to promptly.

In a small herd, a man can often cope with special situations and inconveniences but, in large herds, such hazards quickly find expression in lowered efficiency.

Emphasis on Observation

Careful and critical observation of each and every cow all

the time is an essential feature of good cowmanship.

It must be recognised that observation is a part of every operation and that it deserves constant emphasis. It involves all the senses that the operator possesses—sight, touch, smell, hearing and, rarely, taste. Using all the senses calls for special effort and makes real demands on the operator's reserve of energy.

It is unlikely during periods of heavy physical work, or of intense concentration on executing a routine, that the operator will have enough energy available to register adequately the whole situation mentally. In such circumstances, observation periods have to be consciously programmed into the routine of the operator.

Intelligent men become skilled in observing large groups of animals. They recognise abnormality in the herd pattern itself and from that point their attention is focused on the identification of abnormal individuals, if indeed individuals are the cause. This may not be the case. Some external or circumstantial feature may be responsible and it is perhaps responsible for affecting all or most of the animals.

Animal Clocks

Cattle are creatures that settle down to routines. Any deviation is liable to cause primary disturbance in the behaviour pattern and this can, in turn, result in accident or lost production.

However, during the dry period, and to a lesser extent during the first twelve weeks of lactation, cows accept changes in routine without much apparent loss of production compared with what happens if the change is made after the 12th and before the 30th week of lactation.

A herdsman must try to carry through the day's routine with precision so far as sequence of operation and timing are concerned. Otherwise, the cows become uneasy and restless.

If changes have to be made, they should be carried through gradually and persuasively over an adequate period.

It is always more desirable to place a cow in the position in which only one course of action is available to her (which almost inevitably she eventually discovers and follows of her own volition) than to force her to adopt a course of action which appears to be one of many available to her.

Herd Movement

A herd moving of its own volition and at its own speed tends to string out in a long line and the individuals follow each other along a path or paths.

Horned cattle follow each other at distances apart which are greater than is the case with polls and they tend to develop more distinct paths. Polls wear wider paths with less distinct edges because the herd moves in more compact mass.

When a herd is driven, the pressure behind is felt more keenly by those at the rear and the herd assumes the shape of a triangle. As the driving pressure is increased, the triangle becomes wider-based and more densely packed.

Under extreme rear-pressure the herd becomes a line of cattle all abreast of each other and ultimately the driving force —man, dog or machine—breaks through the rank and the herd is divided into two ranks each speeding in opposite directions at right angles to the former line of flight. This happens particularly with horned cattle. Polled cattle, in contrast, tend to run into an impenetrable mass when hustled along.

Driving cattle involves hazards when the rate of progress exceeds the natural, easy pace of the slower members of the herd. The shape of the triangle reveals the degree of danger to the herd at a casual glance.

Driving too quickly has bad effects on the cattle and the sward.

It is easy to damage feet, legs and udder and to upset the cow before milking. The sward is excessively cut up by a herd driven at speed because it is on account of the foot's sharp backward thrust that the turf is penetrated and the mud flies.

A well-routined herd usually comes well-enough when called —a long line, each animal at its own pace and in its proper place.

At grazing time, the herd is a natural unit and there is a tendency for all to graze and then for all to rest. The herd moves around large fields in a group with leaders and followers clearly identifiable to the observant.

Long water troughs are preferred for large herds because drinking, like grazing, is a herd routine and its members prefer to drink as a herd rather than to do so on an individual basis throughout the twenty-four hours.

In small paddocks, however, where stocking rate exceeds 60 per hectare, there is a tendency for the herd to saturate the whole area for movement to be individual and ideally minimal. There are doubts as to the wisdom of increasing stocking density above 74 per hectare because it can lead to disturbance and adversely-affected cow performance and grassland management.

Grazing behind an electric fence, however, can be practised with quite large herds packed quite closely together, as is explained presently.

The Aggression Scale

The behaviour of the herd is partly based on the fact that some cows are more aggressive than others and seek strongly to behave as they wish. The aggression scale has become termed 'the bunt order' and the leaders (the brave) are at the top of the scale and the followers (the timid) at the bottom.

In mass-handling cows, especially in loose-housing, regard should be paid to the facts of the aggression scale. It explains the milking order, the activities at the silage-face, the trough-losses and many accidents.

The role of the herdsman is basically to reduce the aggression scale as far as is possible. Good feeding arrangements help enormously—'bunt order' is highly operative at feed time—and every effort must be made to ensure that each cow gets her quota of the several feeds.

When each cow was tied up by the neck each cow was positively provided for. Self-feed systems have their hazards for nervous cows at the bottom of the bunt order.

In modern times, the risks are diminished for these animals in several ways—ad lib supplies of silage at a restricted face for 24 hours, generous trough-space, locking feed-stall yokes, tombstone feeders, oblique rails dividing the trough into individual feed spaces, generous supply of medium-palatable bulk-feed, etc.

Good resting arrangements are also important in order to

keep the cows clean, and to enable them to chew the cud, rest and process the feed efficiently. Cow cubicles are proving to be highly successful in that they give the individual cow its necessary opportunity to rest in comfort and security.

The Act of Aggression

Nature provided horned cattle with organs of aggression and defence—horns.

The vulnerable part of the beast is the soft under-belly and this is the area that is attacked by the aggressor. The reaction of the attacked animal is to wheel its rear end away from the attack and to present its own horns in its own defence. Hence the battle of strength, the victor pushing the vanquished so that it presents its vulnerable area for attack. The vanquished either quickly wheels into attack again or escapes at speed.

In nature, the battle is not to kill but to establish social rights of precedence in the herd and consequently it is rare for females to complete the act by tearing open the belly.

Naturally-polled beef cattle behave differently. They attack usually by burrowing charges to the shoulder, heart and belly of the opponent. Dis-horned dairy cattle occasionally exhibit these traits but with less force and fury.

In handling cattle it is well to remember these basic facts that motivate members of the herd and the fact that the vulnerable area is below the flank.

It explains the success of the cubicle which provides flank-protection and also the success of the electric fence where all the animals are orientated in such a way that those grazing realise that they are secure against flank attack. Similarly, in feed lines with tombstones which anchor, even if temporarily, the cows in the line.

The bunt order regulates the lives of all members of the herd all the time, especially under intensive conditions. Each one has its approximate place in the social hierarchy.

The most aggressive claims the best place, at rest, when grazing, in the feed line, at the trough and at the water point. The most timid spend much of their time escaping challenges and behaving in such a way that they could not be construed by others as to be mistakenly misunderstood to be offering a challenge.

For this reason, if one enters a crowded field or a yard of cattle, the most aggressive ones find security in the remoter parts and because they occupy those areas the weaker ones are pressurised forward towards you because they fear you less than the aggressors. Thus you see the weaker ones in these circumstances.

If, however, a herd is inspected from a feedway you see the strong ones in the yard because they hold their advantageous positions near the troughs quite satisfied that you cannot reach them. In the same way, anyone inspecting cattle in a field from an adjacent roadway, or even from within a vehicle in the field, is likely to see the best and strongest and most aggressively curious members of the herd.

The Cow's Senses

While we have developed great dependence on the sense of sight, it is probable that cattle rely as much on smell and hearing as they do on sight.

For this reason, cattle should always be allowed to smell their way into places—slowly the first time. Once they have become familiar with the smells, noises and light values of a situation their fears largely abate and, on future occasions, they enter confidently and without trouble.

It is good sense, therefore, to give heifers plenty of time and to coax them gently to enter the parlour the first few times. Brutally driving them in leads them afterwards to associate the process with fear and injury.

Teaching to Lead

Very young calves are easily tied up with light halters to rings in a wall. This is a good way to prevent navel-sucking and ear-sucking after they have been fed. Such calves are said to be 'imprinted' because they never fail to respond to the halter when they grow up and this is the easiest and safest way to teach cattle to lead.

Herd and Cow Control

The inexperienced man is always *running after* cattle and he is usually trying to recover a situation from which they have escaped. He fails to recognise what they are going to do until it is too late and consequently he starts preventive action too late.

He will gradually learn to watch their eyes all the time. In short, he concentrates on the job in hand and every move of the beasts is as quickly countered. Cutting horses from Canada and dogs can do it. Man can do it even better.

When moving dry stock, it is advantageous to induce a slight degree of hysteria to bind the members together as a unit. A stick is useful in as much as it is an extension of the arm for guiding purposes only. It should always be kept down as sticks held high seem to alarm cattle.

If it is necessary to take one beast home, for example, it usually pays to take several to accompany it. This is particularly necessary with polls that show a decided tendency to break back to join the rest. The cattle left behind should remain in a pen, if possible, until the ones removed are well out of sight and beyond hearing range.

Some skilled people can cut out individual beasts and take them where they want to without fuss or excitement. They work quietly and quickly.

Probably they have achieved this peculiar capacity because they are able to see themselves as the cattle see them. They estimate and anticipate all the time. They wear cloth caps and have no flapping coat tails. They speak in quiet coaxing terms and avoid shouts, screams, squeals and all high-pitched notes.

CHAPTER XI

YOUR INDUSTRY

IT may help to complete the picture of herdsmanship if I attempt, in this last chapter, to give an overall view of the great industry in which you play such an important part.

The present cattle population (June, 1977) in the United Kingdom is around 13,854 millions, compared to 14,717 millions in 1975. The national dairy herd has fluctuated between 3-3½ millions, 3,264 millions in 1977.

Bulls being reared for service have shown a further decline, mainly as a result of the increasing use of AI, so that bulls in use or being reared for service now represent less than 0.7 per cent of the cattle population. Over 65 per cent of dairy cows are now bred artificially by AI.

Breed Pattern

The breed pattern in the national dairy herd of England and Wales has changed somewhat over the same period, as the table on the next page shows.

The significant feature of this table is the decline in the dual-purpose breeds compared to the dairy breeds. The position of the Friesian breed is unique in that besides its high milk producing ability, steer calves of the breed have made a tremendous contribution to our beef supplies.

DAIRY HERD BREED DISTRIBUTION
Cows and heifers in milk and cows in calf

Breed Type	*England and Wales*		*Scotland*		*Northern Ireland*	
	1965	1973-4*	1969	1975	1969	1976
	Cows of Each Breed as % of All Breeds					
Ayrshire	15·7	3·6	68·6	47·0	6·7	2·4
Friesian	64·2	81·0	18·2	27·9	69·5	91·3
Guernsey	5·7	2·8	1·0 (Guernsey and Jersey)	0·7	If any, included in Crosses and Others	
Jersey	4·3	2·2				
Dairy Shorthorn	6·3	0·9	See Crosses and Others	See Crosses and Others	21·4	5·4
Red Poll	0·4	—			If any included, in Crosses and Others	
South Devon	0·8	0.2				
Welsh Black	0·1	—	—	Others		
Others	0·1	—	0·3	0·2		
Total Beef	0·9	9·3 (Total Beef to Crosses and Others)	‡	‡		
Unspecified	1·5		See Crosses and Others			
Friesian/Ayrshire X	†		11·6	24·1		
Crosses and Others	†		0·3	0·1	2·4	0·9
Total	100·0	100·0	100·0	100·0	100·0	100·0
No. of cows ('000)	2,647	3,086	317	314	207	258

**Results of a farm survey based on the Board's Permanent Producer Sample and carried out in the winter of 1973-4. The 9·3% "Others" etc., includes 8·2% cows in mixed breed dairy herds. The total number of cows is as estimated from the Sample.*

†Crosses included with the breeds they most closely resemble.

‡Beef cows have been excluded.

Imported Breeds of Cattle

The importation of highly-bred Friesians from Holland in 1936 and 1946 set the scene for further introduction of European, North American and New Zealand cattle with the object of improving our cattle stocks and adapting them to modern requirements. Since then Canadian Holsteins and Guernseys have been imported from Canada and Jerseys from New Zealand. In order to secure polled blood for developing polled strains of various breeds importations of Shorthorns, Herefords and North Devons were made from the United States, Canada and New Zealand.

The Ayrshire Society imported semen from Finland and the Shorthorn Society imported Red Danish blood from Denmark.

However, the Diseases of Animals Act, 1950, made importation of cattle difficult and the Ministry required to be convinced that it would be of benefit to British breeds and desired by sufficient livestock producers who were interested in cross-breeding.

It was on these terms that 31 Charolais bulls were imported in 1961 after a Committee, appointed in 1959 by the Minister to advise regarding an experimental importation, did so in 1960.

The Charolais is a large white beef breed with a high growth rate potential from Central France. Early trials, which involved the various AI centres in testing the 31 bulls on dairy cows, suggested advantages in food conversion, killing-out percentage and carcase grading.

The anticipated calving difficulties, associated with the breed in France, were confirmed here. However, the trouble was more common in certain lines of bulls and when these bulls were used on cattle of particular breeds.

In 1965 after the trials were reported it was decided to make further importations and particularly to establish the breed here. These were as follows:

19	bulls and	200	females in	1965
10	,, ,,	31	,, ,,	1966
26	,, ,,	243	,, ,,	1967
2	,, ,,	2	,, ,,	1969
25	,, ,,	319	,, ,,	1970
12	,, ,,	439	,, ,,	1971
150	both bulls and females together			1972

In March, 1970, it was announced that imports of both Limousin and Simmental cattle would be permitted because they warranted trial here.

It was in 1971 that 25 bulls and 155 females of the Limousin breed were imported. The Limousin came in, from around Limoges in France, on the ground of their potential as a hardy, deep-fawn, large-bodied beef breed. They are said to have impressive uniformity of type, fleshing ability, high killing-out percentage and good carcase quality. These cattle are reputed to have a high fertility rate and the advantage of easy calving, early maturity and longevity in the females. Purebred, they are good suckler cows and, used as a crossing sire, they impart quick-growing, early-maturing carcases.

In 1970 47 bulls and 487 females of the Simmental breed and in 1971 nine bulls and 62 females were imported. The Simmental, originating in Switzerland but now spread over central Europe especially Germany and France, is a multi-purpose breed-beef, milk and draught. It produces deep lean meat and a milk yield not much less than that of a Friesian. Its usefulness as a crossing sire, it is anticipated, lies in its potential for improving the milking capacity of cows in our suckler herds.

The Shorthorn Society wished to use the Meuse-Rhine-Ijssel breed in their breed improvement programme and they imported six bulls in 1970. The Dutch red-and-white breed, the Meuse-Rhine-Ijssel (MRI abbreviated) is a good dual-purpose animal and a good producer of milk of high butterfat content.

In August, 1971, the Minister announced the Government's intention to liberalise the policy on the import of breeding cattle and semen. In consequence of this proclaimed change of policy already arrangements are going ahead to import 150 head each of two breeds, Maine-Anjou and Blonde D'Aquitaine (1972).

The Maine-Anjou is a dual-purpose French breed of Shorthorn type, very large with rapid growth and calculated to achieve improved beef production when crossed with the Shorthorn breeds.

The Blonde D'Aquitaine is a recently formed breed in Southern France and it has been used there for producing veal calves which reputedly make rapid growth.

This liberalised policy has resulted in applications to import many breeds, allegedly as many as 22, including the Chianina (a beef breed in Italy) and the Gelbvieh (a yellow beef breed in West Germany).

It appears that at the time, post-war, when the Hereford and the Friesian were achieving dominant positions among our own traditional breeds, other forces were at work which led to thinking about the introduction of new types of cattle which had great size, leanness, vigour and growth rate in their favour. First, the basic idea was to adjust and improve our own breeds, but latterly it is becoming more likely that pure breeds of some of the imported cattle will be established and that bulls of these breeds will play their part in dairy herds siring some of the beef calves surplus to the requirements of the dairy herd.

The Hereford sire is firmly established in that role, too, and the crossbred females make very good suckler cows and it is likely that some of the bulls of the new breeds will sire beef calves from these suckler dams on better land where growth potential can be expressed.

Beef from Dairy Herds

The policy of more beef from dairy herds has been encouraged by the calf rearing subsidy* on calves of beef type—£11.25 for steers and £9 for heifers (excluding the Guernsey, Jersey, Ayrshire and Friesian breeds) reduced in 1973 to £8.50 and £6.50 for steers and heifers respectively, born after 16th April, 1973. This arrangement was revised again in early 1974 so that calves born after 30th October, 1973, and before 1st January, 1975, should be eligible for an additional sum of £10 in both cases, i.e. £18.50 and £16.50. The trend has also been assisted by disease eradication from dairy herds—by March 1961 the whole of Britain was fully attested. Longer herd life with dairy cows means fewer replacement heifers need be reared and in 1962 a further step forward in the control of contagious abortion was taken with the introduction of a free vaccination scheme (see page 103).

The incidence of disease in the national dairy herd of Britain, as revealed by a survey in 1957/58 carried out by the Ministry of Agriculture, gave the following information:

ESTIMATED INCIDENCE OF THE VARIOUS DISEASES IN THE NATIONAL DAIRY HERD

	Estimated incidence %	*Estimated number of animals affected*
1. Diseases associated with parturition:— difficult calving, stillbirth, abortion, retained placentas, other general diseases	11	315,600
2. Infectious diseases:— mastitis (acute, mild, summer), winter scours, Johne's disease, others	15	409,800
3. Non-infectious diseases:— milk fever, acetonaemia, grass tetany, bloat, traumatic conditions, lameness, other conditions	17	478,300

Note: The national dairy herd in Britain at the time of the survey was approximately 3 million cows.

The non-infectious diseases and parturition problems are a challenge to our management, as has been emphasised in the text of this book; the infectious diseases are being tackled by

*Calf subsidy was terminated in April 1978

research workers and veterinary practitioners in the dairy farming field.

Tackling Disease

Antibiotic treatment and better milking machine operation are helping in the control of mastitis. With Johne's disease there is hope of a suitable vaccination technique being developed as a result of long-term field trials now being conducted by the Veterinary Officers of the Ministry of Agriculture.

In beef cattle, the chief sources of loss are at calving and in the early stages of calfhood. The mortality in young calves is still too high, particularly in calves exposed to starvation and exposure during transit from farm to farm, especially if the period of colostrum feeding has been too short.

Vaccination against husk has offered a means of reducing losses in young cattle despite higher rates of stocking which accompany intensification of calf rearing, as has occurred on many farms.

Continued Expansion

The general picture presented by the cattle industry today, is of continued expansion. Dairy herds are getting larger from an average of 25 cows in 1965 to 45 cows by 1976 and yield per cow has risen to 4,320 litres in 1976 compared with 3,580 litres 10 years earlier.

Output per man has risen appreciably also, input being on average 124 labour hours per cow in 1960/61 compared to 65.3 hours in 1968/69. This has largely been achieved by modernisation of fixed equipment under the Farm Improvement Schemes and the adoption of labour-saving methods of milking (parlour systems) and housing (yarding) of cows.

In beef production, winter feeding systems have also been streamlined to save labour and the emphasis today is on the early finishing of beef animals as baby beef (12-14 months old) or beeflings (18-20 months). The latter, making wider use of grassland as grazing, is probably the type of beef most likely to meet economic conditions in the future.

Technical Report No 4 of the Beef Recording Association sets out clearly the methods and business aspects of "Commercial Beef Production".

Five Systems of Farming

On the individual farm the choice of milk or beef production or both is determined by individual circumstances, mostly of an economic character. We can broadly define the following five systems of farming with cattle:

1. *Store cattle rearing in association with sheep farming*

On land usually over 240 m (often in conjunction with some rough grazing), low rental value (hence maintenance costs of the breeding herd are low), cows live out all the year round, generally calving in January/May; characteristic breeds are Highland, Luing, Galloway, Hereford and Welsh Black or their crosses with Aberdeen Angus and Shorthorn, often mated with Aberdeen Angus or Hereford bulls.

Each cow rears her own calf which is weaned in the autumn and generally sold off as a suckled calf, large sales of which take place annually.

These suckler herds require the minimum of labour and capital equipment, a system which fits in well with sheep farming in the fuller utilisation of hill grazings. The breeding cows qualify for a compensatory allowance which varies with the value of the "Green Pound"—about £30 per cow.

Where progress in land reclamation and improvement or increase in farm size by amalgamation of holdings have justified increasing the herd size in hill country, many farmers have taken advantage of grant-aided schemes to erect modern labour-saving buildings. It is an advantage when large beef herds are housed to have them calve in September/October, especially in the modern cubicle houses. The risk of injuring the young calves is much reduced if they are two months old, when they are housed in November/December in a modern intensive layout.

2. *Store cattle rearing with sheep and some milk selling*

Characteristic of much upland farming, but on lower, kinder land than system 1, the system has been intensified by keeping cows of more "milky" characteristics such as North Devon, Sussex, or the Cumberland or Lincoln Red Shorthorns, and calf rearing by more intensive methods involving rearing by nurse cows suckling several calves, surplus milk being sold.

Because of the easier provision of winter keep the calves are not so often sold as sucklers, but kept on until more mature and sold as up to 18-month-old stores for fattening on the lowlands.

Both of these systems illustrate the joint enterprise nature of beef production in that the rearing of stores on the relatively lower productivity hill lands is linked with the fattening of these stores on lowland farms where the higher level of nutrition required for fattening relative to rearing can be fully met from grazing or arable land.

In these areas the Hereford (bull) x Friesian (cow) is a popular choice and autumn calving is usually preferred.

3. *Beef production in conjunction with dairy or arable farming*

Characteristic of the larger lowland farms, particularly in the arable areas of England and Scotland. Steer calves of dual-purpose breeds like Red Poll and Dairy Shorthorn, and from dairy herds of British Friesian cattle, can be reared and in some cases fattened out on the same farm.

In other cases colour-marking bulls are being used in conjunction with the above pure breeds (and even in Ayrshire herds) on a proportion of the dairy stock.

Rearing in these herds is either by nurse cows or by pail feeding.

4. *Specialised dairy farming*

Increasing dependence on milk-selling is characteristic of small farms, particularly those below 40 hectares where the quicker turnover and greater monetary output per hectare from milk production compared with any form of beef production dictates the cattle policy.

Many of these small farms may rear no stock and those that do tend to concentrate on dairy stock for herd replacement purposes only though Brucellosis accreditation experience has indicated the clear advantage of rearing calves born on the farm for herd replenishment.

The main effective contribution these small farms can make to beef production is in the supply of calves to the larger arable farms in group 3. Somerset, Cheshire, Staffordshire, Derbyshire and East Lancashire are typically intensive dairying areas "exporting" calves for rearing elsewhere.

5. *Specialised beef production*

The maintenance of pedigree herds of beef cattle for stud purposes is a feature of a restricted number of lowland farms. Such herds meet the demand for pedigree bulls for export and for crossing purposes.

Such stud farms demand a high level of feeding and general management, and are often located on productive land with well-equipped farm buildings. Similar farms may, in the arable areas, engage in winter beef production or the wintering of beef stores, particularly where sugarbeet is extensively grown.

Influences on Farm Production

The more productive the land for arable cropping, the less likely is rearing of any form to be undertaken. Instead relatively mature beef stores around 18 months old are bought in for finishing.

Similarly, on first-class fattening pastures rearing is rarely seen. The seasonal nature of grass fattening presents a wintering problem unless cattle requiring less than six months' fattening are purchased each spring—as is customary in most grazing areas with little or no arable land.

Such then is the general pattern of the cattle industry, from the purely rearing hill or upland areas to the fattening and dairying areas on the more fertile lowlands, but the stratification within the industry is far from rigidly defined in terms of hill or valley land.

Milk production, particularly in its economic advantages to the small farmer and the opportunity it offers for judicious use of purchased feedingstuffs, is determined not so much by land fertility as by the presence or absence of suitable buildings, houses, water and electricity supplies and so on.

Good Stockmanship

No matter how well planned a farming system may be, its success ultimately depends on how successful is the execution of that plan, and in cattle farming good stockmanship reaps a rich dividend.

With increasing responsibilities which are the lot of the herdsman today, his education and devotion to the job are so important, and it is hoped that by reading this book the aim to do even better will be encouraged. It is also my hope that such men will receive their just reward in terms of good wages and modern housing—they deserve it.

APPENDICES

Appendix I

TABLES OF FOOD COMPOSITION*

	Dry Matter Content g/kg	*Metabolisable Energy*	*Digestible Crude Protein g/kg DM*
Roots			
Carrots	130	12·8	62
Mangolds, intermediate	120	12·4	58
Parsnips	150	13·3	67
Potatoes	210	12·5	47
Swede Turnip	120	12·8	91
Turnip	90	11·2	73
Leaves of Roots			
Sugar beet tops	160	9·9	88
Turnip leaves	120	9·2	130
Other Green Foods			
Cabbage, drumhead	110	10·4	100
Cabbage, open leaved	150	10·8	115
Kale, thousandhead	160	11·1	106
Kale, marrow stem (unthinned)	140	11·0	123
Rape	140	9·5	144
Cereals			
Barley in flower	250	10·0	46
Maize	190	8·8	53
Oats in flower	230	8·6	61
Grasses			
Pasture grass, close grazing: Non-rotational	200	12·1	225
Rotational, with three weekly intervals	200	12·1	185
Pasture grass, extensive grazing Spring value, running off during summer	200	10·0	124
Winter pasturage (after close grazing allowing free growth from end of July to December)	200	9·7	101
Ryegrass perennial post flowering	250	8·4	72
Green Legumes			
Lucerne, in bud	220	9·4	164
Sainfoin, early flower	230	10·3	143
Vetches, in flower	180	8·6	123

**Selection of feeds taken from Technical Bulletin No.* 33 *published by the Ministry of Agriculture, Fisheries and Food.*

Appendix I—continued

	Dry Matter Content g/kg	*Metabolisable Energy*	*Digestible Crude Protein g/kg DM*
Silage — Clamp			
Alsike	250	8·6	80
Clover (Red)	220	8·8	135
Grass (very high digestibility)	200	10·2	116
Grass (high digestibility)	200	9·3	107
Grass (moderate digestibility)	200	8·8	102
Grass (low digestibility)	200	7·6	98
Lucerne	250	8·5	113
Maize	210	10·8	70
Silage — Tower			
Barley (whole crop)	400	9·6	50
Grass (very high digestibility)	400	10·4	121
Grass (high digestibility)	400	9·3	87
ROUGHAGES			
Hay			
Clover, red very good	850	9·6	128
Grass (very high digestibility)	850	10·1	90
Grass (high digestibility)	850	9·0	58
Grass (moderate digestibility)	850	8·4	39
Grass (low digestibility)	850	7·5	45
Grass (very low digestibility)	850	8·2	166
Dried Grasses and Legumes			
Grass, very leafy	900	10·8	113
Grass, leafy	900	10·6	136
Grass, early flower	900	9·7	97
Lucerne, just in bud	900	9·4	174
Straws and Chaff			
Barley straw, Spring	860	7·3	9
Barley straw, Winter	860	5·8	8
Oat straw, Spring	860	6·7	11
CONCENTRATES			
Grain and Seeds — Cereals			
Barley	860	13·7	82
Maize	860	14·2	78
Oats	860	11·5	84
Rye	860	14·0	110
Wheat	860	14·0	105
Legumes			
Beans, field spring	860	12·8	248
Beans, field winter	860	12·8	209
Peas	860	13·4	225

Appendix 1—continued

	Dry Matter Content g/kg	*Metabolisable Energy*	*Digestible Crude Protein g/kg DM*
Oil Seeds			
Linseed	900	19·3	208
Oil Cakes and meals			
Coconut cake	900	13·0	184
Coconut cake meal	900	12·7	174
Cotton cake, Brazilian	900	8·9	234
Cotton cake, decorticated	900	12·3	393
Ground nut cake, decorticated	900	12·9	449
Ground nut cake, undecorticated	900	11·4	310
Ground nut meal, decorticated	900	11·7	491
Ground nut meal, undecorticated extracted	900	9·2	316
Linseed cake, English made	900	13·4	286
Linseed meal, extracted	900	11·9	348
Palm kernel cake	900	12·8	196
Palm kernel meal, extracted	900	12·2	204
Soya bean cake	900	13·3	454
Soya bean meal, extracted	900	12·3	453
FEEDINGSTUFFS OF ANIMAL ORIGIN			
Blood meal	900	13·2	848
Fish meal, white	900	11·1	631
Pure meat meal	900	16·3	753
Feeding meat meal (high fat)	900	13·3	624
Feeding meat meal (low fat)	900	11·1	631
Meat and Bone meal (high protein)	900	9·7	465
Meat and Bone meal (medium protein)	900	7·9	411
Mill, cows whole	128	20·2	250
By-products			
Barley, brewers' grains, fresh	220	10·0	149
Barley, brewers' grains, ensiled	280	10·0	149
Barley, brewers' grains, dried	900	10·3	145
Barley, distillers' grains, fresh	250	11·8	237
Barley, distillers' grains, dried	900	12·1	214
Barley malt culms	900	11·2	222
Maize, flaked	900	15·0	106
Maize germ meal, high fat	900	14·9	116
Maize germ meal, low fat	900	13·2	90
Maize, gluten meal	900	14·2	339
Potato pulp (dry)	860	10·8	0
Rice sludge, dried	860	13·6	250
Sugar beet pulp, pressed	180	12·7	66
Sugar beet pulp, dried	900	12·7	59
Sugar beet pulp, molassed	900	12·2	61
Wheat feeds, middlings	880	11·9	129
Yeast, dried	900	11·7	381

Appendix II

USEFUL CALCULATION FACTS

Stacks of Hay and Straw

First, work out the cubic capacity of the stack.

For rectangular stacks multiply length by breadth (in yards) and multiply that figure by height to eaves. Then add cubic capacity of roof—obtained by multiplying the length by breadth again and multiplying that figure by half the height of the roof, measured perpendicularly from eaves to ridge.

For round stacks, square the circumference, multiply by the height to the eaves and multiply the result by 0.08. Then add the cubic capacity of the roof—obtained by multiplying the area at the eaves (0.08 $\times$ circumference squared) by one-third of the height of the roof, measured perpendicularly from the eaves to peak. Make all the above measurements in yards.

When the total cubic capacity is known, tonnage can be worked out from the following tables:

AVERAGE NUMBER OF CUBIC METRES (M^3) PER TONNE OF HAY

Condition of stack	*Square stacks*	*Round stacks*
Not well settled	9	10
Fairly well settled	7·5	8
Very well settled	6	7

AVERAGE NUMBER OF M^3 PER TONNE OF STRAW

Wheat straw	13–15
Oat straw	9–17
Barley straw	15–17

Tower Silos

The weight of silage per cubic foot is variable, depending on the depth of silage, diameter of silo, fineness of cutting, packing, etc., but, on an average 1 m^3 weighs 664 kg.

Pit Silos

Work out the cubic capacity in yards—multiply length by breadth by height, taking average measurements—then reckon that each 1 m^3 weighs 664 kg.

Appendix II—continued

Roots

The average weight per m^3 of roots in clamp is:

Fodder-beet	536 kg
Swedes	536 kg
Mangolds	571 kg

TABLE FOR USE IN CALCULATING WINTER RATIONS

Quantity available (kg)	*Amounts available per head over—*			
	3 *months* (90 *days*)	4 *months* (120 *days*)	5 *months* (150 *days*)	6 *months* (180 *days*)
100	1·1	0·8	0·7	0·6
200	2·2	1·7	1·3	1·1
300	3·3	2·5	2·0	1·7
400	4·4	3·3	2·7	2·2
500	5·6	4·2	3·3	2·8
1,000	11·1	8·3	6·6	5·6
2,000	22·2	16·7	13·3	11·1
3,000	33·3	25·0	20·0	16·7
4,000	44·4	33·3	26·7	22·2

Appendix

BREEDING TABLE

JAN Serv-ed on	Due on	FEB Serv-ed on	Due on	MAR Serv-ed on	Due on	APR Serv-ed on	Due on	MAY Serv-ed on	Due on	JUNE Serv-ed on	Due on
1	OCT 11	1	NOV 11	1	DEC 9	1	JAN 9	1	FEB 8	1	MAR 11
2	12	2	12	2	10	2	10	2	9	2	12
3	13	3	13	3	11	3	11	3	10	3	13
4	14	4	14	4	12	4	12	4	11	4	14
5	15	5	15	5	13	5	13	5	12	5	15
6	16	6	16	6	14	6	14	6	13	6	16
7	17	7	17	7	15	7	15	7	14	7	17
8	18	8	18	8	16	8	16	8	15	8	18
9	19	9	19	9	17	9	17	9	16	9	19
10	20	10	20	10	18	10	18	10	17	10	20
11	21	11	21	11	19	11	19	11	18	11	21
12	22	12	22	12	20	12	20	12	19	12	22
13	23	13	23	13	21	13	21	13	20	13	23
14	24	14	24	14	22	14	22	14	21	14	24
15	25	15	25	15	23	15	23	15	22	15	25
16	26	16	26	16	24	16	24	16	23	16	26
17	27	17	27	17	25	17	25	17	24	17	27
18	28	18	28	18	26	18	26	18	25	18	28
19	29	19	29	19	27	19	27	19	26	19	29
20	30	20	30	20	28	20	28	20	27	20	30
21	31	21	DEC 1	21	29	21	29	21	28	21	31
22	NOV 1	22	2	22	30	22	30	22	MAR 1	22	APR 1
23	2	23	3	23	31	23	31	23	2	23	2
24	3	24	4	24	JAN 1	24	FEB 1	24	3	24	3
25	4	25	5	25	2	25	2	25	4	25	4
26	5	26	6	26	3	26	3	26	5	26	5
27	6	27	7	27	4	27	4	27	6	27	6
28	7	28	8	28	5	28	5	28	7	28	7
29	8			29	6	29	6	29	8	29	8
30	9			30	7	30	7	30	9	30	9
31	10			31	8			31	10		

In Leap Year one day must be added to birth

III

FOR COWS

JULY Served on	Due on	AUG Served on	Due on	SEPT Served on	Due on	OCT Served on	Due on	NOV Served on	Due on	DEC Served on	Due on
1	APR 10	1	MAY 11	1	JUNE 11	1	JULY 11	1	AUG 11	1	SEPT 10
2	11	2	12	2	12	2	12	2	12	2	11
3	12	3	13	3	13	3	13	3	13	3	12
4	13	4	14	4	14	4	14	4	14	4	13
5	14	5	15	5	15	5	15	5	15	5	14
6	15	6	16	6	16	6	16	6	16	6	15
7	16	7	17	7	17	7	17	7	17	7	16
8	17	8	18	8	18	8	18	8	18	8	17
9	18	9	19	9	19	9	19	9	19	9	18
10	19	10	20	10	20	10	20	10	20	10	19
11	20	11	21	11	21	11	21	11	21	11	20
12	21	12	22	12	22	12	22	12	22	12	21
13	22	13	23	13	23	13	23	13	23	13	22
14	23	14	24	14	24	14	24	14	24	14	23
15	24	15	25	15	25	15	25	15	25	15	24
16	25	16	26	16	26	16	26	16	26	16	25
17	26	17	27	17	27	17	27	17	27	17	26
18	27	18	28	18	28	18	28	18	28	18	27
19	28	19	29	19	29	19	29	19	29	19	28
20	29	20	30	20	30	20	30	20	30	20	29
21	30	21	31	21	JULY 1	21	31	21	31	21	30
22	MAY 1	22	JUNE 1	22	2	22	AUG 1	22	SEPT 1	22	OCT 1
23	2	23	2	23	3	23	2	23	2	23	2
24	3	24	3	24	4	24	3	24	3	24	3
25	4	23	4	25	5	25	4	25	4	25	4
26	5	26	5	26	6	26	5	26	5	26	5
27	6	27	6	27	7	27	6	27	6	27	6
28	7	28	7	28	8	28	7	28	7	28	7
29	8	29	8	29	9	29	8	29	8	29	8
30	9	30	9	30	10	30	9	30	9	30	9
31	10	31	10			31	10			31	10

date of animals served after February 28th.

INDEX